INTRODUCTION TO BACTERIA

for students in the biological sciences

INTRODUCTION TO BACTERIA

for students in the biological sciences

Paul Singleton
Diana Sainsbury

JOHN WILEY & SONS
Chichester · New York · Brisbane · Toronto

Copyright © 1981 by John Wiley & Sons Ltd.

British Library Cataloguing in Publication Data:

Singleton, Paul
  Introduction to bacteria.
  1. Micro-organisms
  I. Title   II. Sainsbury, Diana
  576   QR41.2

ISBN 0 471 10034 X Cloth
ISBN 0 471 10035 8 Paper

Typeset by Photo-Graphics, Yarcombe, Devon, England
and printed in the United States of America
by Vail-Ballou Press, Inc., Binghamton, N.Y.

Contents

Preface

The microbiologist is well served by an abundance of large textbooks, each covering the many aspects of his subject. However, for the non-specialist — requiring only the essentials of one or two of these aspects — such books are daunting in their sheer bulk and mass of detail. Hence the need has long been recognized for smaller, introductory texts dealing individually with the various groups of microorganisms, and there is currently available a variety of elementary texts on fungi, viruses, protozoa and algae. However, to date there is no corresponding text for the bacteria. This is a surprising omission in view of the importance of bacteria in so many branches of biological science — genetics, biochemistry, ecology, medicine, plant pathology, food science, biotechnology. In biotechnology, for example, the rapid growth of techniques for genetic manipulation (using bacteria as tools) — and the tremendous potential of such techniques in research and industry — means that a knowledge of the rudiments of bacteriology is likely to become even more important to biologists in the future.

In writing this book we have tried to give a concise but none the less thorough account of bacteriology for students taking any of a wide range of courses in the biological sciences. Bearing in mind the differing academic backgrounds that these students will have, and the varied nature of the courses they will be following, we have tried to present a broadly-based treatment and to avoid bias towards any particular aspect of the subject. For this reason the book should also be useful as a primer for students embarking on more specialized courses in bacteriology or microbiology.

Paul Singleton & Diana Sainsbury
Exeter, January 1981.

1 The bacteria: an introduction

What are bacteria?

Bacteria are minute organisms which occur almost everywhere. Sometimes they reveal their presence — wounds may 'go septic', milk 'sours', meat 'putrefies' — but usually we are unaware of them because their activities are less obvious and because they are so small. Indeed, the very existence of bacteria was unknown until the development of the microscope.

In most cases an individual bacterium is a single cell. However, the bacterial cell is unlike the cells of most other organisms. Bacterial cells are of the *prokaryotic* type, while those of most other organisms — including higher animals and plants — are of the *eukaryotic* type. Prokaryotic and eukaryotic cells differ in many important respects. For example, in a eukaryotic cell the chromosomes (the thread-like bodies of genetic material) are enclosed by a double membrane to form a distinct structure, the nucleus. In the prokaryotic cell there is no nuclear membrane — the (usually single) chromosome being in direct contact with the cytoplasm of the cell. The major differences between eukaryotic and prokaryotic cells are given in Table 1-1; many of the prokaryotic features mentioned in this table will be considered in more detail in later chapters.

Another group of prokaryotic organisms, formerly called the 'blue-green algae', are sometimes considered to be bacteria and have been renamed *cyanobacteria*. These organisms do have many features in common with bacteria, but they also show some important differences; we shall use the term 'bacteria' in a sense which does *not* include the cyanobacteria.

Table 1-1 Eukaryotic and prokaryotic cells: some distinguishing features.

Eukaryotic cells	Prokaryotic cells
The chromosomes are enclosed within a double sac-like membrane.	The chromosome is not enclosed within a sac-like membrane.
The chromosomes are complex in structure and their DNA is usually associated with proteins called histones.	The chromosome is relatively simple in structure and histones are never present.
Cell division occurs by mitosis or meiosis.	Mitosis and meiosis never occur.
The cell wall, when present, contains compounds such as cellulose or chitin, but never peptidoglycan.	The cell wall, when present, usually contains peptidoglycan but never cellulose or chitin.
Mitochondria are generally present, and chloroplasts occur in photosynthetic cells.	Mitochondria and chloroplasts are never present.
Cells contain ribosomes of two types: a larger type in the cytoplasm and a smaller type in mitochondria and chloroplasts.	Cells contain only one type of ribosome which resembles the smaller of the two types of eukaryotic ribosome.
Flagella, when present, have a complex structure.	Flagella, when present, have a relatively simple structure.

Why study bacteria?

At first sight it may seem that the conquest of disease is the most important reason for studying bacteria. It is well known that some bacteria can cause disease — although it should be borne in mind that many diseases are caused not by bacteria but by viruses, fungi, or protozoa. Diseases which are caused by bacteria include, for example, typhoid and syphilis in man, anthrax, brucellosis and tuberculosis in both man and animals, and certain types of wilt and soft rot in plants. Many such diseases have been conquered or controlled largely as a result of studies and experimental work carried out on the causal agents by medical, veterinary and agricultural bacteriologists.

Important though they are, the disease-causing bacteria represent only a very small proportion of the bacteria as a whole. Most bacteria do little or no harm, and indeed, many are positively useful to man.

Some, for example, help in our fight against disease by producing a number of important antibiotics. Many bacteria are important because their activities are essential to the re-cycling of matter upon which, ultimately, all life depends. For example, some species of soil bacteria bring about chemical changes which are essential steps in the nitrogen cycle, and such changes have important effects on the nitrogen content of the soil. Since certain forms of nitrogen (nitrate, ammonia) are necessary for plant growth, an understanding of this type of bacterial activity is essential for better management of land and crops — so vital to the survival of our ever-expanding population.

Surprisingly, perhaps, bacteria also make a significant contribution in our food industry. We usually think of bacteria as a nuisance where food is concerned, causing spoilage and 'food poisoning', but particular species of bacteria are actually used in the production of some types of food. For instance, the manufacture of dairy products such as butter, cheese, and yoghurt depends on the ability of certain bacteria to convert milk sugar to lactic acid; furthermore, the characteristic flavours of these products often owe much to other compounds produced by the bacteria during the process of manufacture. Bacteria are also employed in the manufacture of certain vitamins (eg. vitamin B_{12}) and amino acids.

This is by no means a complete account of the importance of bacteria in our everyday lives. However, from what has been said it should be clear that the more we can learn about these active but unseen organisms the more effectively we can minimize their harmful effects and exploit their useful abilities.

Classifying and naming bacteria

How is one species of bacterium distinguished from another, and how are bacteria classified? Different species of bacteria may differ from one another in shape and size, in their chemical activities and nutritional requirements, in the physical conditions under which they can grow, and in their reactions to certain dyes. There are other differences — some of which are explored in later chapters — but, in general, features of the type listed above are those used in classification. Such a system of classification is *not* intended to indicate evolutionary relationships. *Species* of bacteria which resemble one another are grouped into a *genus*, and genera with similar characteristics may be grouped into a *family*; families may in turn be grouped into an *order*. A bacterial species can be subdivided into *strains* — organisms which conform to the same species definition but which have various minor differences.

As in the case of animals and plants, each species of bacterium is given an internationally accepted name in the form of a Latin binomial. In a binomial the first name is that of the genus to which the organism belongs, and the second is the 'specific epithet' which refers to one particular species within that genus. A binomial is usually printed in italics or, if handwritten, underlined once; a genus name always begins with a capital letter. Examples of bacterial binomials include *Escherichia coli, Salmonella typhi, Nitrobacter winogradskyi*. The genus name may be abbreviated — eg. *Escherichia coli* may be abbreviated to *E. coli*; however, this should be done only when the full genus name has previously been mentioned so that the meaning of the abbreviation is always clear. The names of orders and families of bacteria are not necessarily written in italics but do have a capital initial letter; these names have standardized endings, the name of an order always ending in *ales* (eg. Actinomycetales) and the name of a family always ending in *aceae* (eg. Enterobacteriaceae).

2 The bacterial cell

Shapes, sizes and arrangements

Bacterial cells vary widely in shape, according to species. Spherical cells are called *cocci* (singular: coccus), while elongated, rod-shaped cells are called *bacilli* (singular: bacillus) or simply *rods*. This does not mean that all cells called cocci are necessarily precisely spherical, or that all bacilli are precisely the same shape. There are, for example, cocci which are more or less kidney-shaped, and bacilli which taper at each end (*fusiform* bacilli) or which are curved (*vibrios*). Ovoid cells, intermediate in shape between cocci and bacilli, are called *coccobacilli*. There are also two types of spirally-shaped cells: those which are rigid (*spirilla,* singular: spirillum) and those which are flexible (*spirochaetes*). Finally, there are the actinomycetes — a group of bacteria most of which, like many fungi, grow in the form of fine threads called *hyphae* (singular: hypha). Examples of some bacterial cell types are given in Fig. 2-1. (It will be noticed that some of the names given to cell types double as genus names — eg. *Bacillus, Spirillum, Vibrio;* care should be taken not to confuse the two. A bacillus, for example, may or may not belong to the genus *Bacillus.*)

While the cells of most species of bacteria are more or less uniform in shape, there are some species in which the cells typically show some variation in form — even within a single population. Such variation is known as *pleomorphism.* Cell shape can also vary to some extent according to the conditions under which the organism is growing.

The dimensions of a bacterial cell are usually measured in *micrometres,* μm (= microns, μ) 1μm = $1/1000$ of a millimetre. Bacteria range in size from 'midgets' of 0.2μm or less (eg. certain species of *Mycoplasma*) to 'giants' measuring up to 500μm in length (eg. species of *Spirochaeta*). However, these are extreme cases; in the majority of species the maximum dimension of a cell lies within the range 1-10μm.

6

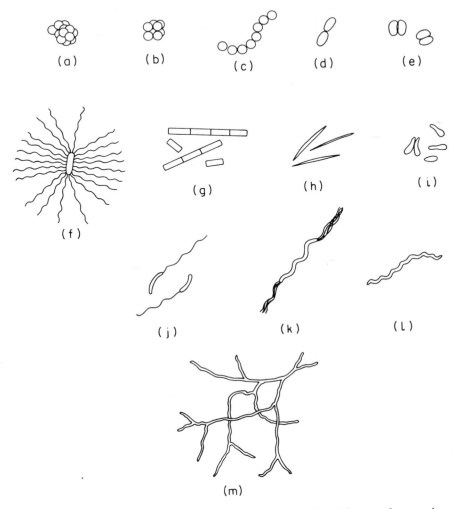

Fig. 2-1 Some shapes and arrangements of bacterial cells, with named examples. (Not drawn to scale.) (a) Uniform spherical cells (cocci), in irregular clusters: *Staphylococcus aureus*. (b) Cocci, in regular packets of eight: *Sarcina ventriculi*. (c) Cocci, in chains: *Streptococcus pyogenes*. (d) Slightly elongated cocci, in pairs (diplococci): *Streptococcus pneumoniae*. (e) Diplococci in which each cell is flattened or slightly concave on the side next to its neighbour: *Neisseria gonorrhoeae*. (f) Rod-shaped cell (bacillus), peritrichously flagellated*: *Escherichia coli*. (g) Bacilli, individually and in chains: *Bacillus anthracis*. (h) Bacilli with tapered ends (fusiform bacilli): *Fusobacterium nucleatum*. (i) Irregularly shaped (pleomorphic) cells: *Corynebacterium diphtheriae*. (j) Curved bacilli (vibrios), monotrichously flagellated*: *Vibrio cholerae*. (k) Rigid spiral cells (spirilla), lophotrichously flagellated*: *Spirillum volutans*. (l) Flexible spiral cells (spirochaetes): *Treponema pallidum*. (m) Thin branched filaments (hyphae): *Streptomyces* species.
*For details of flagella see later.

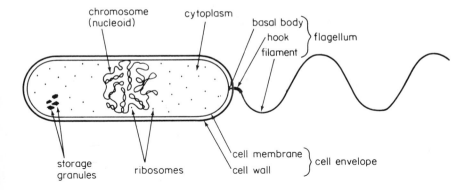

Fig. 2-2 A generalized bacterial cell.

(Note that the smallest bacteria are invisible under the ordinary light microscope which has a limit of resolution of about 0.2μm.)

In some bacteria the cells occur in characteristic groupings of two or more cells. Depending on species, cells may occur in pairs, in irregular clusters, in chains, in packets of eight, etc; Fig. 2-1 gives some examples.

The bacterial cell: a closer look

Fig. 2-2 shows, in diagrammatic form, a generalized bacterial cell. Not all bacteria have all of the features shown in this diagram; for example, some have no flagellum (or numerous flagella) while a few have no cell wall. On the other hand, certain bacteria may have features in addition to those shown in Fig. 2-2, and some of these features will be considered later in the chapter.

In many bacteria the cell wall (Fig. 2-2) can be removed from a living cell under experimental conditions to form a 'naked' cell called a *protoplast*. Although very delicate, the protoplast can survive (under laboratory conditions) and can carry out many of the processes of a normal living cell. In a protoplast only the cell membrane separates the cytoplasm from the environment, so that this structure must clearly play a vital role in regulating the passage of substances into and out of the cytoplasm.

The cell membrane The cell membrane (also called the cytoplasmic membrane or plasma membrane) is composed primarily of lipids and

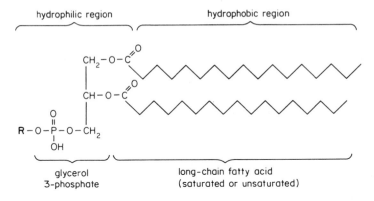

Fig. 2-3   A glycerophospholipid. R may be any of a number of substituents — eg. ethanolamine ($^+NH_3.CH_2.CH_2-$).

proteins. The lipids are mainly glycerophospholipids, ie. fatty acid esters of glycerol 3-phosphate — see Fig. 2-3. It can be seen from Fig. 2-3 that a phospholipid molecule has a compact hydrophilic (water-attracting) 'head' region and a long hydrophobic (water-repellent) 'tail' region. In the cell membrane the phospholipids are arranged in two layers (a *bilayer*) in such a way that the hydrophilic groups occur on either side of the bilayer while the hydrophobic chains occur within it. Thus a sheet is formed in which a hydrophobic layer is sandwiched between two hydrophilic layers (Fig. 2-4). However, this bilayer is far from being a rigid structure. The lipid molecules are actually in a fluid state: they can move freely through the membrane in a sideways direction, but they cannot 'flip' from one side of the bilayer to the other.

Some of the proteins in the membrane are associated with one side or the other of the bilayer, while others are embedded within it —

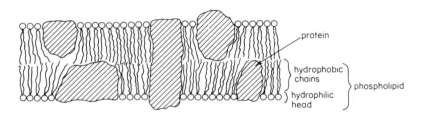

Fig. 2-4   The cell membrane: a diagrammatic illustration of a cross-section, showing the arrangement of lipids and proteins within the bilayer.

sometimes extending across its width (Fig. 2-4). The proteins appear not to be purely structural but are involved in the various functions of the membrane.

The cell membrane is not freely permeable to most molecules, although certain small uncharged or hydrophobic molecules can pass through it more or less unhindered. How then do other molecules such as nutrients enter the cell? Such molecules must be *transported* across the membrane by special mechanisms, and in most cases each type of molecule has its own specific transport system. Many of the membrane proteins are concerned with these transport systems — and the mechanisms involved are varied and often complex. Some require an expenditure of energy by the cell (Chapter 5) and allow the cell to accumulate a particular substance to a concentration which far exceeds that in the surrounding medium.

As well as transport, a number of other important processes take place in the cell membrane — including, for example, the bacterial respiratory system (Chapter 5). The cell membrane also contains enzymes involved in the synthesis of new cell wall and cell membrane components.

The cell interior The bacterial cytoplasm is generally regarded as a kind of 'soup' containing nutrients, waste products, enzymes etc in solution. There seems to be no structure equivalent to the eukaryotic endoplasmic reticulum, and little is known about the way in which the cytoplasm may be organized.

Suspended within the cytoplasm is the genetic material of the bacterium, the *chromosome,* which consists essentially of a closed loop of DNA. (We shall be looking at the structure of DNA in Chapter 6.) It has been estimated that the chromosome of *Escherichia coli* is 1 millimetre or more in length — ie. 500-1000 times longer than the organism itself — so that clearly it must be extensively folded to fit within the cell. The chromosome is attached by at least one point to a site on the cell membrane.

The cytoplasm also contains numerous small rounded bodies called *ribosomes,* and these are the sites of protein manufacture in the cell. Ribosomes are made of protein and RNA (a polymer similar to DNA — see Chapter 6). In bacteria the ribosomes are approximately 0.02μm in diameter and are smaller than those found in the cytoplasm of eukaryotic cells — as indicated by the rate at which they sediment in a high-speed centrifuge. Bacterial ribosomes have a sedimentation rate of 70S as compared with 80S for ribosomes of the eukaryotic cytoplasm.

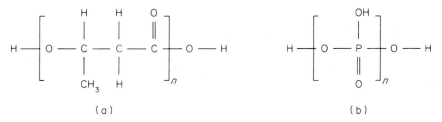

(a)   (b)

Fig. 2-5 Two common bacterial storage compounds: (a) poly- β-hydroxybutyrate, and (b) polyphosphate (volutin).

(S stands for Svedberg unit, which is simply a measure of the sedimentation rate.) Each 70S ribosome consists of two subunits which, when separated, sediment at rates of 30S and 50S. We shall consider the role of ribosomes in protein synthesis in Chapter 6.

In many bacteria the cytoplasm may contain various types of insoluble granules — some of which function as reserve supplies of food. A common storage compound is a polymer called *poly-β-hydroxybutyrate* (Fig. 2-5(a)); the cell can draw on its internal store of poly-β-hydroxybutyrate for supplies of carbon and energy when external sources become depleted. Another common storage compound is *polyphosphate* (Fig. 2-5(b)), sometimes called *volutin*. Polyphosphate granules have an unusual response to certain stains — adopting a colour different from that of the stain used. This phenomenon is known as metachromasy and hence polyphosphate granules are sometimes called *metachromatic granules*. A few bacteria (including some photosynthetic species) may contain granules of elemental sulphur. Some aquatic spirilla are curious in that they contain particles of magnetite (magnetic iron oxide) which cause them to behave like miniature magnets!

Specialized structures may be found in the cytoplasm of certain bacteria. For example, the photosynthetic bacteria and cyanobacteria have special internal membranes (see Chapter 5) which contain the components of their photosynthetic apparatus. Some aquatic bacteria and cyanobacteria have gas-filled, protein-walled vesicles (*gas vesicles*) which give buoyancy to the cells. In many bacteria membranous structures called *mesosomes* have frequently been observed under the electron microscope; however, no function for these membranes has been clearly established, and it is possible that they may be artefacts derived from the cell membrane by certain experimental procedures.

Cytoplasm contains many substances in solution. If a protoplast is suspended in a medium more dilute than its cytoplasm, water will pass through the cell membrane into the cytoplasm by osmosis, and the protoplast will swell and eventually burst — an event known as *osmotic lysis*. Of course, in an intact bacterial cell this does not happen because, in most species, the protoplast is protected by a tough outer layer called the cell wall.

The cell wall is a rigid layer which encloses the delicate protoplast and protects it from osmotic lysis and from mechanical damage. The wall also maintains the shape of the cell: protoplasts are spherical,

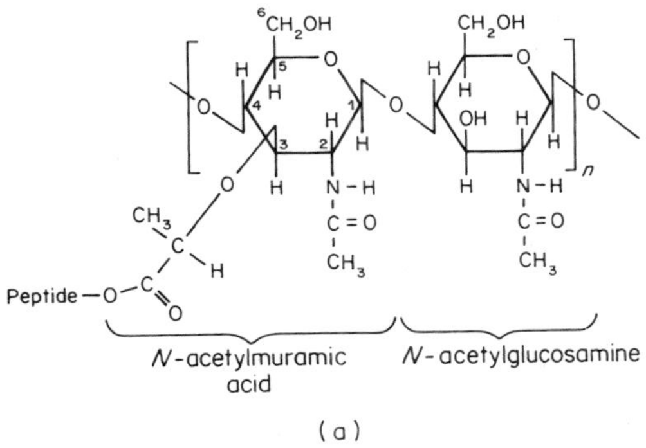

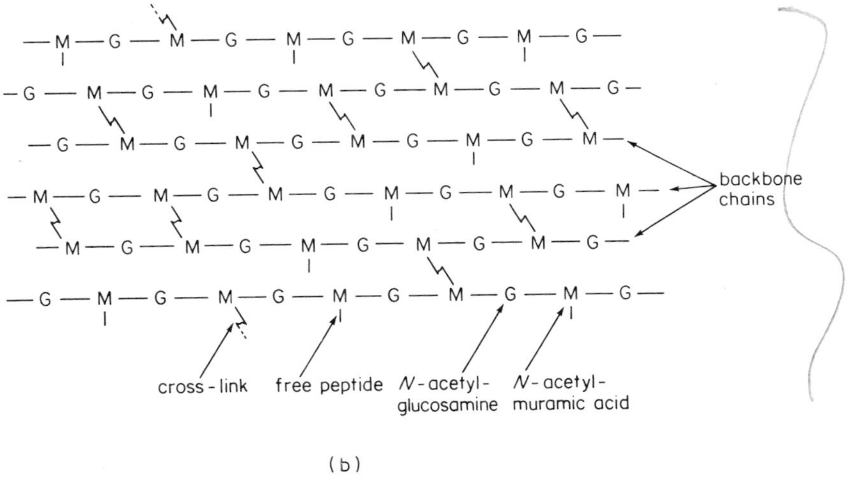

(a)

(b)

Fig. 2-6 Peptidoglycan. (a) The structure of the polysaccharide backbone chains. (b) A diagrammatic representation of the structure of a cross-linked peptidoglycan molecule.

regardless of the shape of the cells from which they were derived. The component generally thought to be responsible for the rigidity of the wall is a polymer unique to prokaryotic cells: *peptidoglycan* (also called murein or mucopeptide). The main components of peptidoglycan are unbranched polysaccharide backbone chains — each of which consists of two alternating amino-sugars: N-acetylglucosamine and N-acetylmuramic acid (Fig. 2-6). Attached to some or all of the muramic acid residues is a short chain of amino acids (ie. a peptide). Some of the peptides on one backbone chain are linked to peptides on adjacent chains — either directly or through an additional peptide bridge; such cross-links between many backbone chains (shown diagrammatically in Fig. 2-6(b)) give rise to a giant hollow molecule — sometimes referred to as the *sacculus* — which completely encloses the protoplast.

Nearly all bacterial cell walls contain peptidoglycan, but not all bacterial cell walls resemble one another in other respects. In fact there are two main types of cell wall which differ from each other in composition and structure and in their reaction to certain dyes. When heat-killed bacterial cells are stained with a dye such as crystal violet and then treated with iodine, a dye-iodine complex forms inside the cells. If these cells are now treated with an organic solvent, such as acetone or ethanol, one of two results will generally be obtained. Either the organic solvent will not extract the dye-iodine complex from the cells (which therefore remain stained), or the complex will be completely washed out of the cells (which are thus decolorized). This differential staining procedure reflects a fundamental difference in the composition and structure of the two main types of cell wall; it was discovered in the 1880s by the Danish scientist Christian Gram and so is known as the *Gram stain*. Cells which are *not* decolorized by the organic solvent are said to be *Gram-positive,* while those which *are* decolorized are called *Gram-negative*. (Practical details of the Gram stain are given in Chapter 11.)

The Gram-positive type of cell wall is the simpler of the two types and generally has a uniform appearance when examined under the electron microscope: Fig. 2-7(a). In this type of cell wall peptidoglycan is a major constituent and is interspersed with other components — especially *teichoic acids*. (Teichoic acids are polymers of glycerol phosphate or ribitol phosphate and may contain amino acids and/or sugars.)

The Gram-negative cell wall is considerably thinner than the Gram-positive type and appears as a distinctly layered structure under the

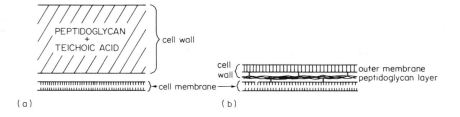

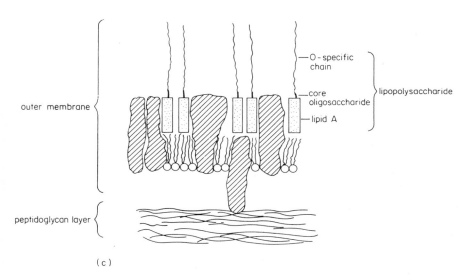

Fig. 2-7 Bacterial cell walls: Gram-positive and Gram-negative types. (a) A Gram-positive cell envelope (cell wall + cell membrane). (b) A Gram-negative cell envelope. (c) The Gram-negative cell wall showing the structure of the outer membrane; phospholipids and proteins are depicted as in Fig. 2-4.

electron microscope: Fig. 2-7(b). Peptidoglycan forms a relatively small proportion of the whole and occurs as the innermost layer of the wall, immediately surrounding the cell membrane. The peptidoglycan is not interspersed with other components although it is usually linked to the outer layers of the wall. These outer layers are composed of phospholipids, proteins, lipoproteins, and a polymer unique to Gram-negative cell walls called *lipopolysaccharide*. Lipopolysaccharide is a major component of the wall; it consists of three parts: a complex lipid called *lipid A*, the *core oligosaccharide*, and the *O-specific chain* — a polysaccharide which consists of repeating oligosaccharide subunits. The lipopolysaccharide, protein and lipid components together make up a membrane-type bilayer, not unlike that of the cell membrane, called the *outer membrane.* In the outer membrane the lipopolysaccharide occurs in the outermost layer, with the O-specific polysaccharide chains extending outwards; most or all of the phospholipid occurs in the innermost layer — Fig. 2-7(c). As in the cell membrane, proteins may be attached to or embedded within the outer membrane.

The bacterial cell wall should not be regarded as an inert box which simply contains the living cell. Recent research has shown that it has many functions in addition to protecting the cell from damage. For example, certain wall components — especially the teichoic acids in Gram-positive bacteria — bind divalent cations such as Mg^{2+} and regulate their flow through the wall; ions such as these are required for many reactions in the cell. The wall also acts as a kind of molecular sieve which blocks the passage of larger molecules (eg. some antibiotics). In the outer membrane of Gram-negative bacteria certain proteins (called *porins*) are involved in the formation of pores which allow free passage across the outer membrane of certain small hydrophilic molecules such as amino acids and sugars. The outer membrane also contains transport systems which are specific for certain other nutrient molecules, eg. vitamin B_{12}.

Capsules and slime layers The nature of the bacterial cell *surface* is important since it is the region of contact between a cell and its environment. In some bacteria the cell wall itself forms the outermost layer, but in others an additional layer may be present. Some bacteria secrete a polysaccharide which adheres to the cell to form an external layer called a *capsule;* a few species produce a polypeptide capsule. Other bacteria produce a more fluid type of secretion which adheres less firmly to the cell and tends to diffuse into the surrounding medium; this is called a *slime layer.* Capsules and slime layers may enable bacteria to stick to surfaces, and may also afford the cells some protection against desiccation and damage. In some pathogenic species the capsule plays an important role in the disease-causing process — see Chapter 9.

Flagella and motility Many bacteria are *motile* — ie. they can actively propel themselves through a fluid medium. Most motile bacteria owe their motility to one or more thread-like structures called *flagella* (singular: flagellum) — see Fig. 2-2. Depending on species, a cell may have a single flagellum at one end (a *monotrichous* arrangement) or at each end (an *amphitrichous* arrangement), it may have a

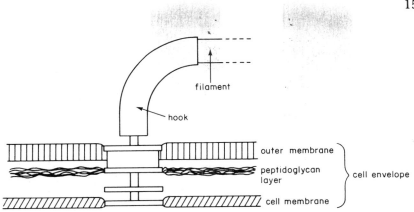

Fig. 2-8 The attachment of a flagellum to the cell envelope in a Gram-negative bacterium such as *Escherichia coli*. The system of four parallel rings embedded in the cell envelope constitutes the basal body of the flagellum; in Gram-positive bacteria only two rings constitute the basal body.

tuft of flagella at one or both ends (*lophotrichous* arrangements), or flagella may be distributed all over the cell surface (a *peritrichous* arrangement): see Fig. 2-1. A flagellum is made up of three distinct parts: the basal body, the hook region, and the filament or shaft. The basal body is a complex structure which anchors the flagellum in the cell envelope; it is connected to the filament through the protein hook region (Fig. 2-8). The filament, which is made of a protein called flagellin, forms a more or less rigid spiral and is typically longer than the cell itself. In some bacteria (eg. *Vibrio* species) the filament has an outer sheath which is continuous with the outer membrane of the cell wall. Motility is achieved by the *rotation* of the flagellum from the basal body — ie. the flagellum acts rather like the propeller of a boat. In peritrichously flagellated cells the flagella bunch together and the rotations of the individual flagella appear to be co-ordinated.

Escherichia coli, a peritrichously flagellated bacterium, moves in a series of straight lines interrupted at intervals by a tumbling motion which is caused by a reversal in the direction of rotation of the flagella. Tumbling results in a *random* change in the direction of motion of the cell. However, the cell can adjust the *overall* direction of its motion by changing the frequency with which it tumbles when moving in a particular direction. For example, a cell can, in effect, swim towards higher concentrations of certain chemicals (called attractants) by tumbling less frequently when moving up the concentration gradient, but at normal frequency when moving in other directions. Similarly, the cell can move away from noxious chemicals (repellants) by tumbling less frequently when travelling down a concentration gradient of the repellent. The ability to respond to chemicals in this way (a pheno-menon known as *chemotaxis*) can be advantageous to the bacterium, allowing it to swim towards sources of nutrients and away from harmful substances.

Flagella are not the only means by which bacteria achieve motility. A number of species of bacteria — and some cyanobacteria — are capable of a *gliding* motion when in contact with a solid surface. Gliding does not appear to involve any change in cell shape, neither has any locomotory organelle been detected; the mechanism of this type of motility remains a mystery. The spirochaetes (Chapter 12) have no external flagella but are capable of characteristic twisting and jack-knifing movements; such movements are brought about by structures called *periplasmic fibrils* (or *axial filaments*) which are sandwiched between the peptidoglycan layer and the outer membrane of the cell wall. Periplasmic fibrils appear to resemble normal bacterial flagella in structure.

Fimbriae and pili *Fimbriae* (singular: fimbria) are straight, hair-like protein appendages which are found mainly on Gram-negative bacteria. They may be distributed all over the cell surface or may occur in a tuft at one particular site on the cell. Fimbriae (also called common fimbriae or common pili) appear to play an important part in enabling the bacteria to stick to surfaces, to other types of cell (including animal cells), and to each other.

Pili (singular: pilus) are also hair-like protein appendages which occur on certain Gram-negative cells; they have a special role in a process called conjugation in which genetic material is passed from one cell to another (see Chapter 6 for details). Pili are sometimes called sex pili or, confusingly, sex fimbriae.

3 Growth and differentiation

GROWTH

Growth in a bacterial cell involves a _co-ordinated_ increase in the masses of its component parts. It is not simply an increase in cell mass since this could be due, for example, to the accumulation of a storage compound within the cell. In a bacterial cell growth usually leads to the division of that cell into two similar or identical daughter cells; thus growth and reproduction are very closely linked in bacteria, and the term 'growth' is often used to cover both processes.

Conditions for growth in bacteria

Bacteria can grow only when environmental conditions are suitable; if conditions are not suitable growth may occur very slowly or not at all, or the bacteria may die, depending on the nature of the species and on the conditions.

Perhaps the most obvious requirement for growth is a supply of suitable nutrients. As a group the bacteria can use a vast range of compounds as nutrients — although an individual species may be able to use only a very limited range. Even substances such as sugars and amino acids (widely used as nutrients) cannot be used by all species of bacteria. The concentration of a nutrient is also important: if present in too high a concentration a nutrient may actually inhibit growth.

Many types of bacteria can grow only when oxygen is present, and such bacteria are called _aerobes_ — sometimes 'strict aerobes' or 'obligate aerobes' to emphasize their absolute requirement for oxygen. However, some bacteria — the _anaerobes_ — can grow only when oxygen is absent. Yet others, the _facultative anaerobes,_ can grow in the presence or absence of oxygen, while a few species (the _microaero-_

philes) grow best when the concentration of oxygen is lower than that normally present in air.

The presence of water is a very important factor in bacterial growth. This is hardly surprising since water makes up 80% or more of the mass of a bacterium, and nutrients and waste products must enter and leave the cell in solution. Hence, bacteria can grow only in or on materials which have a high content of free, available water. (Not all of the water actually present in a given material is necessarily available for bacterial growth; for example, some may be 'tied up' by hydrophilic gels or by ions in solution.)

Generally, growth in a given species of bacterium occurs most rapidly at a particular temperature: the *optimum growth temperature*. The rate of growth tails off at temperatures above and below the optimum, and for each species there are maximum and minimum temperatures beyond which growth does not occur. Bacteria whose optimum growth temperature is higher than about 45°C are called *thermophiles*. Such organisms occur, for example, in compost heaps and in the hot springs of Yellowstone National Park (USA) — natural waters whose temperature (80-85°C) would be lethal for most organisms. (Note that some bacteria can tolerate high temperatures even though their optimum growth temperature is below 45°C; such bacteria are said to be *thermoduric*.) Bacteria which grow optimally in the temperature range 20-45°C are called *mesophiles;* they occur in a wide range of habitats and include most of the species which cause disease in man and other animals. Bacteria which can grow at very low temperatures, eg. 0°C and below, are called *psychrophiles;* these organisms inhabit polar seas and can also be found in refrigerated foods. (The term 'psychrophile' means different things to different people. One definition which is becoming widely accepted is as follows: a psychrophile is an organism which has an optimum growth temperature of 15°C or below, a maximum temperature for growth of 20°C or below, and a minimum temperature for growth of 0°C or below. Organisms which do not fit this definition but which can nevertheless grow at low temperatures are sometimes called *psychrotrophs.*)

Most bacteria grow best at or near pH 7 (neutral), and the majority cannot grow under strongly acidic or strongly alkaline conditions. However, certain bacteria not only tolerate but even 'prefer' extremes of pH; for example, *Thiobacillus thiooxidans* (an organism which actually produces sulphuric acid) thrives in the pH range 0-3 and is therefore said to be *acidophilic*. Organisms which are not acidophilic but which can nevertheless tolerate low pH are said to be *aciduric*.

All bacteria require the presence of certain inorganic ions (such as Mg^{2+}, Fe^{3+}, Cl^-) in low concentrations, but high concentrations of ions are usually inhibitory to growth. However, the *halophilic* bacteria — found eg. in the sea and in salt lakes — need relatively high concentrations of salt (sodium chloride) for survival and growth, and the extreme halophiles may require salt concentrations in excess of 20%.

Of course, no factor which influences bacterial growth operates in isolation, and an alteration in one factor may enhance or reduce the effects of another. For example, the maximum temperature at which a bacterium can grow may well be reduced if the acidity of its environment is increased.

Growth in a single cell

We have already seen that growth in a bacterial cell involves a co-ordinated increase in its component parts. Nevertheless, all the different parts of a cell do not increase at the same time, but rather do so in a definite sequence. In a bacillus such as *Escherichia coli*, for example, there is an initial increase in cell length which necessarily involves extension of the cell wall and cell membrane. This means that new components must somehow be fitted into the wall and membrane without disturbing the integrity of these important structures. Once the cell has reached a certain size the chromosome replicates to form two identical chromosomes (see Chapter 6 for details). At this stage the cell itself begins to divide into two by the development of a partition (*septum*) which is formed by the inward growth of the cell envelope at a site half-way along the length of the cell. The chromosomes are so positioned that when the septum eventually divides the original cell into two, each of the new (daughter) cells contains one chromosome. The daughter cells may separate immediately, or they may remain together so that subsequent cell divisions (in the same plane or in different planes) result in the formation of one of the characteristic groupings mentioned in Chapter 2. The simple division of one cell into two is called *binary fission*.

The sequence of events in which a cell grows and divides into two cells is called the *cell cycle* or *cell division cycle*. The time taken for this cycle of events is called the *doubling time* — the length of which depends on species and on the conditions of growth. Doubling times range from minutes to hours or even days; under optimum conditions in the laboratory *Escherichia coli*, for example, has a doubling time of about 20 minutes.

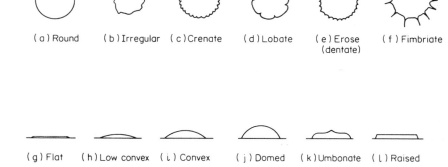

(a) Round   (b) Irregular   (c) Crenate   (d) Lobate   (e) Erose   (f) Fimbriate
                                                          (dentate)

(g) Flat   (h) Low convex   (i) Convex   (j) Domed   (k) Umbonate   (l) Raised

Fig. 3-1 Bacterial colonies: some examples of colony shapes as seen from above (a)-(f) and from one side (g)-(l).

Growth in bacterial populations

Since growth leads to cell division, and since each daughter cell can itself grow and divide, a large population of cells can quickly build up if conditions are favourable. Such a population may develop either on the surface of a solid medium or within a liquid medium ('medium' being the term for any material which can support bacterial growth).

Growth on a solid medium One type of artificial solid medium which is used almost universally in bacteriological laboratories is a jelly-like substance (an *agar gel*) containing nutrients and other ingredients (see Chapter 4). Suppose a single bacterial cell is placed on the surface of such a medium and given all the requirements and conditions necessary for growth and cell division. The cell grows and divides into two cells, and each daughter cell subsequently divides into two. If growth and division continue, the progeny of the original parent cell eventually reach such immense numbers that they form a compact heap of cells that is usually visible to the naked eye; such a mass of cells is termed a *colony*. Typically each species forms colonies of characteristic size, shape (Fig. 3-1), colour, and consistency, although different types of colony may be formed when growth occurs on different types of medium. The size of a colony growing on a given medium may be limited eg. by exhaustion of nutrients or by the build-up of inhibitory waste products (or both), and for similar reasons crowded colonies tend to be smaller than well spaced ones; the rate at which a colony increases in size depends largely on temperature.

Bacteria which produce pigments generally form brightly coloured colonies (eg. red, yellow, violet) while colonies of non-pigmented bacteria are usually grey, whitish or cream-coloured. In consistency a colony may be mucoid (viscous and mucus-like), butyrous (butter-like), friable (crumbly), etc, and its surface may be glossy or dull.

Suppose that, instead of starting with a single cell, we start with a very large number of cells distributed over the surface of the medium; in this case there may be insufficient space for individual colonies to develop, and growth will result in the formation of a continuous layer of bacterial cells which covers the entire surface of the medium. This is referred to as *confluent growth*. Confluent growth can also result when a few cells of a motile bacterium are deposited on the medium; these cells and their progeny can swim through the surface film of moisture and eventually spread to cover the whole surface of the medium.

Growth in a liquid medium Bacteria can move freely through a liquid medium either by diffusion or, in motile species, by active locomotion. Thus, as cells grow and divide, the progeny are commonly dispersed throughout the medium — which usually becomes increasingly turbid (cloudy) as the concentration of cells increases. (Certain bacteria are exceptional in that they tend to congregate in a layer — called a *pellicle* — at the surface of the medium; below the pellicle the medium may be almost free of cells.)

Suppose that a few bacterial cells are introduced into a suitable liquid nutrient medium which is then held at the optimum growth temperature for that species. At regular intervals of time a small volume of the medium can be withdrawn and a count made of the cells it contains (see Chapter 4 for counting methods). By this means we can follow the development of the population — ie. the way in which the number of cells increases with time. By plotting numbers of cells against time we can obtain a *growth curve* which — for a given species growing under given conditions — has a characteristic shape.

When bacteria are introduced into a fresh, unused liquid medium, cell division may not begin immediately — ie. there may be an initial *lag phase* in which little or no cell division occurs. During this period the cells are adapting to their new environment — for example, by making enzymes to utilize the newly available nutrients. The length of the lag phase will depend largely on the conditions under which the cells existed *before* they were introduced into the medium. A long lag phase will often occur if the cells had previously existed under harsh

Table 3-1 Increase in cell numbers with time for *Escherichia coli* growing under optimum conditions in the logarithmic phase.

Time (minutes)	0	20	40	60	80	100	200	....	300
Number of generations (rounds of cell division)	0	1	2	3	4	5	10	....	15
Number of cells	1	2	4	8	16	32	1024	....	32768
Number of cells as a power of 2	2^0	2^1	2^2	2^3	2^4	2^5	.........2^{10}	....	2^{15}

conditions or had been growing in a different type of medium or at a different temperature. The lag phase will be short — or even absent — if the cells had been growing in a similar or identical medium at the same temperature.

Once adapted to the new medium the cells begin to grow and divide at a rate which is maximum for the species under existing conditions; this is the *logarithmic phase* or *exponential phase* of growth. In this phase the overall number of cells doubles at a constant rate, so that cell numbers increase with time in the manner shown in Table 3-1. If we plot a graph of cell numbers against time, the slope of the curve will increase sharply (Fig. 3-2(a)) and the simple arithmetical scale used in the graph will soon be inadequate to deal with the rapidly increasing numbers of cells. There is an easy solution to this problem. Table 3-1 (bottom row) shows that, at any time, the number of cells can be expressed as a power of 2; for example, at 60 minutes there are 8 cells, and this number can be expressed in the form 2^3 — in which 3 is called

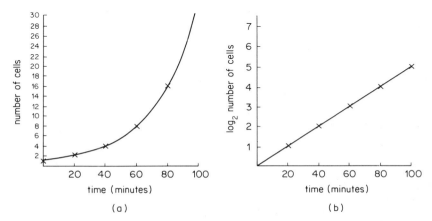

Fig. 3-2 A growth curve in which cell numbers are plotted on (a) an arithmetic scale, and (b) a logarithmic scale.

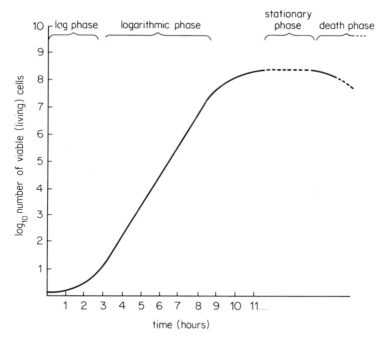

Fig. 3-3 A growth curve constructed for a strain of *Escherichia coli* growing in nutrient broth at 37°C.

the *index*. Instead of plotting cell numbers directly, we can plot the indexes of their corresponding powers of 2, and this gives a straight line graph (Fig. 3-2(b)). Now, if a number is expressed as a power of 2, the index of that power of 2 is actually the *logarithm* (to the base 2) of that number; for example, 8 can be expressed as 2^3, and the index, 3, is the logarithm (to the base 2) of 8. Hence, by plotting these indexes versus time we are plotting the logarithm of the number of cells at each of the times 0, 20, 40, 60... minutes. In such a graph each unit on the $\log_2$ scale represents a doubling in cell numbers, and the doubling time can be read off directly from the time scale of the graph. (It is usually more convenient to use $\log_{10}$ rather than $\log_2$ when constructing a growth curve; $\log_2$ and $\log_{10}$ of any number can be interconverted by using the formula: $\log_{10}N = 0.301 \times \log_2N$. The graph will still be linear — only the slope of the graph changes when the base is changed.)

As they grow, cells use nutrients and produce waste products which accumulate in the medium. Eventually, therefore, growth slows down and stops due either to a lack of nutrients or to the accumulation of waste products (or both); the phase in which there is no overall increase in the number of living cells is called the *stationary phase*. The stationary phase eventually leads into the *death phase* in which the number of living cells in the population progressively decreases (Fig. 3-3).

Diauxic growth If a bacterium growing in a liquid medium is provided with a mixture of two different substrates, it may use one substrate in preference to the other — utilization of the second beginning only after the first has been completely exhausted. For example, if *Escherichia coli* is provided with a mixture of glucose and lactose it uses the glucose first, and will begin to utilize lactose only when all the glucose has been used. During the transition from glucose utilization to lactose utilization growth may slow down or even stop, and the growth curve will therefore show two distinct phases; this pattern of growth is known as *diauxie* (or diauxy).

Synchronous growth In a population of growing cells there are normally slight variations in the growth rate from cell to cell; hence individual members of the population do not all divide at the same instant. However, under special experimental conditions we can obtain a population in which all the cells divide at approximately the same time; this is called *synchronous growth*. In synchronous growth the exponential portion of the growth curve appears as a series of steps — each step representing an abrupt doubling of cell numbers.

Continuous-flow culture When bacteria are grown in a limited volume of liquid medium (ie. as a *batch culture*) the composition of the medium changes continually as nutrients are used up and waste products accumulate. Batch culture is suitable for many types of study, but sometimes it is preferable that cells be grown under constant and controlled conditions. This is achieved by a process called *continuous-flow culture* (or continuous culture) in which the bacteria are grown in a liquid medium contained within an apparatus called a *chemostat*. During growth in the chemostat there is a continual outflow of culture (ie. medium plus cells) and a continual inflow — at the same rate — of fresh, sterile medium; constant agitation is necessary to ensure rapid mixing of the fresh medium with that containing the growing cells. Continuous-flow culture has two important advantages over batch culture. Firstly, the cells are provided with an environment in which fluctuations in the concentrations of nutrients and waste products are kept to a minimum. Secondly, by controlling the rate of flow of medium through the chemostat we can control the rate of growth of the bacteria in the culture.

DIFFERENTIATION

The term differentiation is used here to refer to a process in which one type of cell gives rise to another (different) type of cell. True differentiation occurs in only a limited number of bacteria; a few examples are considered below.

The life-cycle of *Caulobacter*

Caulobacter species are Gram-negative, obligately aerobic bacteria which occur in soil and water. These bacteria form two types of cell which are distinctly different from one another, and the change from one cell type to the other is an essential part of the life-cycle — shown diagrammatically in Fig. 3-4. Note that the motile daughter cell (*swarm cell*) must lose its flagellum and develop a stalk (*prostheca*) before it can divide, and may thus be regarded as immature; the stalked (mature) mother cell can continue to produce swarm cells by growth and division, but cannot itself become a swarm cell. The life-cycle of *Caulobacter* is being intensively studied in the hope that it may throw some light on mechanisms involved in the more complex developmental systems of higher organisms.

The swarming phenomenon in *Proteus*

Two types of cell are also formed (under certain conditions) by each of the species *Proteus mirabilis* and *Proteus vulgaris* — although in this case differentiation is not an essential part of the cell cycle. If cells of *P. mirabilis* or *P. vulgaris* are placed on the surface of a suitable solid medium the first progeny cells to be formed are short, sparsely flagellated bacilli some 2-4μm in length, and these cells form a colony in the usual way. However, after several hours of growth, some of the cells around the edge of the colony undergo a dramatic change — growing to lengths of 20-80μm and developing numerous additional flagella; these cells are called *swarm cells*. Groups of swarm cells migrate outwards to positions a few millimetres from the colony; migration then stops, and each swarm cell divides into several short bacilli by the development of septa at intervals along its length. The

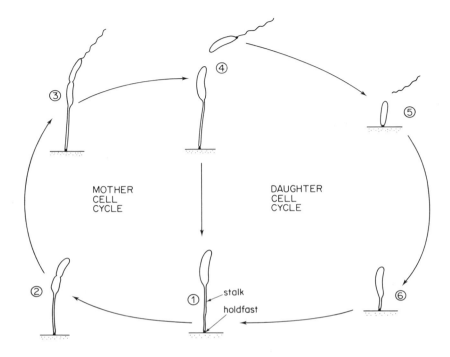

Fig. 3-4 The life-cycle of *Caulobacter*. (1) A mature, stalked cell occurs attached to a surface (or to another cell) by an adhesive holdfast. (2) The stalked cell begins to divide. (3) A flagellum and holdfast develop at the end of the cell opposite the stalk. (4) The cell division is complete and the motile daughter cell (swarm cell) separates from the stalked mother cell. The stalked cell can continue to grow and divide to repeat the cycle. (5) The swarm cell loses its flagellum and becomes attached by its adhesive holdfast. (6) A stalk develops and elongates, and the daughter cell matures into a new stalked mother cell.

short bacilli resemble those of the original colony; they grow and divide normally for several generations, forming a ring of heavy growth which surrounds (and is concentric with) the original colony. Later, another generation of swarm cells is produced at the outer edge of this ring of growth and the cycle is repeated — resulting in another, larger, concentric ring of growth. In this way the entire surface of the medium will eventually be covered by concentric rings of growth. This phenomenon is known as *swarming*.

Resting cells in bacteria

The life-cycle of *Caulobacter* and the swarming phenomenon in *Proteus* are examples in which differentiation and growth are co-ordinated. In some bacteria differentiation can occur as an *alternative* to growth and may result in the formation of a resting cell (either a *spore* or a *cyst*). Resting cells may function as disseminative units and/or as dormant cells which are capable of surviving adverse environmental conditions. Under suitable conditions a spore or cyst *germinates* to give rise to a new vegetative cell.

Bacterial endospores Of the several different types of bacterial spore the one which has been investigated most thoroughly is the

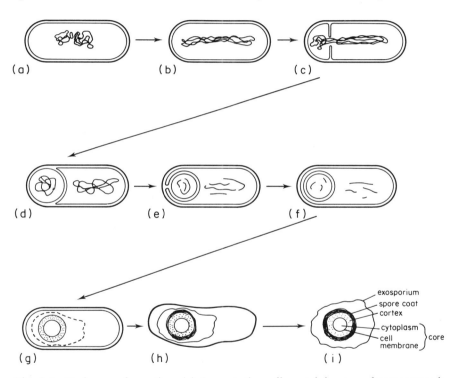

Fig. 3-5 Endospore formation. (a) A vegetative cell containing two chromosomes is about to sporulate. (b) An *axial filament* composed of the two chromosomes develops. (c) The cell membrane grows inwards to form a septum dividing the protoplast into two unequal parts. (d) The septum is complete; the smaller of the two protoplasts will become the endospore and is called the *forespore* or *prespore*. (e) The cell membrane of the larger protoplast invaginates to engulf the forespore. (f) The forespore is completely engulfed and now has two membranes. (g) A layer of modified peptidoglycan is laid down between the two membranes of the forespore to form a rigid layer called the *cortex*. A loose protein envelope called the *exosporium* may begin to develop at about this time. (h) A multilayered protein *spore coat* is deposited as a layer outside the cortex. (i) The completed spore is released by disintegration of the mother cell. (NB. Not all endospore-forming bacteria form an exosporium.)

endospore. Endospores are formed by members of the family Bacillaceae (see Chapter 12) and by certain other bacteria; they are formed mainly as a response to starvation — sporulation occurring when cells are deprived of sources of carbon and nitrogen. An endospore develops *within* a bacterial cell (Fig. 3-5) and appears under the light microscope as a bright (refractile) round or oval structure. The mature endospore exists in a state of dormancy, ie: few — if any — of the chemical reactions characteristic of vegetative cells take place in an endospore. Not only can dormancy persist for long periods of time but the endospore shows a remarkable degree of resistance to unfavourable conditions such as extremes of temperature and pH, desiccation, radiation of various types, many types of chemical agent, and physical damage. The reason for this resistance is not yet clear. Resistance to high temperatures, at least, may be due to the low water content of the endospore as compared with that of the vegetative cell. Characteristically, the core of an endospore contains a high concentration of a substance called dipicolinic acid which occurs complexed with calcium ions; dipicolinic acid was once thought to be responsible for the heat resistance of endospores but it is now believed to be associated with the maintenance of the dormant state.

The *germination* of an endospore is a complex process which is not yet fully understood. It seems to involve three stages: activation, initiation, and outgrowth. *Activation* occurs, for example, when endospores are exposed to heat (at sub-lethal temperatures), to low pH, or to a strong oxidizing agent; the process appears to involve the release of spore enzymes from their dormant state. *Initiation* can be brought about by a variety of conditions — including eg. exposure to certain chemicals (such as glucose or L-alanine in some species) or physical abrasion. In this phase the endospore loses its characteristic properties: dipicolinic acid and calcium ions are released, refractile and resistant properties are lost, and the cortex and spore coat are split or dissolved. In the final stage, *outgrowth,* the conversion of endospore to vegetative cell is completed.

Actinomycete spores Several different types of spores are produced by bacteria of the order Actinomycetales (Chapter 12). Species of *Thermoactinomyces,* for example, form endospores which are essentially similar to those of the Bacillaceae. In members of the family Actinoplanaceae spores are produced within a closed sac (*sporangium*); in most species the spores which are released from the sporangium bear flagella and are motile. Species of *Streptomyces* form

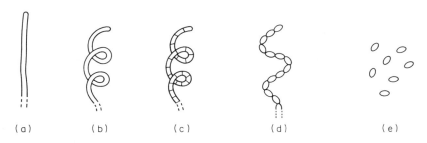

Fig. 3-6 Spore formation in *Streptomyces* species. (a) The tip of a vegetative aerial hypha. (b) The tip of the hypha becomes coiled. (c) Septa develop along the length of the coiled hypha. (d) The walls of the developing spores thicken, and each spore becomes round-ended (ellipsoidal). (e) The spores are released.

spores by the fragmentation of aerial hyphae (ie. hyphae held aloft from the medium): see Fig. 3-6. These spores lack specialized structures such as a cortex and spore coat, but they do have thickened cell walls; they can survive desiccation and exposure to certain chemicals and they also show some resistance to dry heat and radiation. The spores of *Streptomyces* are not completely dormant — although they are less active than vegetative hyphae.

Bacterial cysts A cyst is a cell which, in addition to a thin cell wall, is surrounded by two additional layers of material. The inner layer (the *intine*) is apparently structureless, while the outer layer (the *exine*) is multilayered. Cysts are formed by Gram-negative soil bacteria of the genus *Azotobacter*; *Azotobacter* cysts are resistant to desiccation and radiation but are not significantly more resistant than vegetative cells to heat. Germination occurs when conditions become suitable for growth; unlike endospores, cysts appear not to require activation.

4 Some practical bacteriology

Safety in the laboratory

The bacteriology laboratory differs from most other types of laboratory in that its hazards are less obvious. Accordingly, anyone new to the study of bacteria should be certain that he knows what he is doing *before* doing it, and should constantly remind himself that he is dealing with living organisms — some of which may be able to cause disease. Good bacteriology is safe bacteriology, and it is wise to get to know the safety rules of the laboratory before carrying out any practical work. In particular, the following points deserve special attention:

1. While working in the laboratory wear a clean laboratory coat to protect your clothing. Do not wear the coat outside the laboratory.
2. Put *nothing* into your mouth. It is potentially dangerous to eat, drink or smoke in the laboratory. For pipetting use a rubber bulb (teat) or a mechanical device such as a 'pi-pump'.
3. Keep the bench — and the rest of the laboratory — clean and tidy.
4. Dispose of all contaminated wastes by placing them (not throwing them) into the proper container.
5. Report all accidents and spillages, promptly, to the instructor or demonstrator.
6. Where possible, avoid contaminating the environment with *aerosols* containing live bacteria. An aerosol consists of minute (invisible) particles of liquid or solid dispersed in air; aerosols can form eg. when a bubble bursts, when one liquid is added to another, or when a drop of liquid falls onto a solid surface — in fact, in many normal bacteriological procedures. Particles less than a few micrometres in size can remain suspended in air for considerable periods of time and may be inhaled by anyone in the

vicinity; particles containing live, pathogenic (disease-causing) bacteria are therefore a potential source of infection. Bacteriological work is sometimes carried out in a special cabinet (described later) in order to avoid the risk of infection from aerosols.

7. Always wash your hands thoroughly with soap and warm water before leaving the laboratory.

Bacteriological media

In Chapter 3 we saw that the term *medium* is used to refer to any solid or liquid material on or within which bacteria can be grown. Before use a medium must be *sterile,* ie. it must contain no living organisms. (Methods used for sterilizing media are discussed in Chapter 10.) To grow a particular species of bacterium, the bacteriologist adds to an appropriate sterile medium a small quantity of material which contains living cells of that species; the 'small quantity of material' is known as an *inoculum,* and the process of adding the inoculum to the medium is known as *inoculation.* The inoculated medium is subsequently *incubated,* ie. kept under appropriate conditions of temperature, humidity etc for a suitable period of time. (Incubation is usually carried out in a thermostatically controlled cabinet called an incubator.) During incubation the bacteria grow and divide — giving rise to a *culture;* thus, a culture is a medium containing organisms which have grown (or are growing) on or within that medium. A *pure culture* is one which contains the cells of a single species only, while a *mixed culture* contains the cells of more than one species.

Types of medium No single medium can support the growth of all types of bacteria simply because different species may have different growth requirements. A medium which is suitable only for the growth of nutritionally undemanding species (eg. members of the Enterobacteriaceae — Chapter 12) is referred to as a *basal medium.* Examples of basal media include peptone water and nutrient broth (Table 4-1). (NB. The term 'broth' may refer to any of a variety of liquid media — but commonly refers to nutrient broth or to any liquid medium based on nutrient broth.) Certain bacteria which cannot grow in basal media will do so if the media are supplemented with substances such as egg, serum or blood; media supplemented in this way are referred to as *enriched media.*

Table 4-1   Some common bacteriological media.

Name of Medium	Composition of medium (% w/v in water)
Basal media	
peptone water	peptone (soluble products of protein hydrolysis) 1%; sodium chloride 0.5%
nutrient broth	peptone 1%; sodium chloride 0.5%; beef extract 0.5-1.0%
nutrient agar	nutrient broth gelled with 1.5-2.0% agar
Robertson's cooked meat medium	minced beef heart; beef extract 1%; peptone 1%; sodium chloride 0.5%
Enriched media	
blood agar	nutrient agar containing 5-10% (v/v) citrated or defibrinated blood
chocolate agar	blood agar heated to 70-80°C until the colour changes to chocolate brown
serum agar	nutrient agar containing 5% (v/v) serum
Selective media	
deoxycholate-citrate agar (DCA)	meat extract and peptone 1%; lactose 1%; sodium citrate 1%; ferric citrate 0.1%; sodium deoxycholate 0.5%; neutral red 0.002%; agar 1.5%
MacConkey's broth	peptone 2%; lactose 1%; sodium chloride 0.5%; bile salts (eg. sodium taurocholate) 0.5%; pH indicator bromcresol purple (purple pH 6.8 to yellow pH 5.2) *or* neutral red (yellow pH 8.0 to red pH 6.8)
Enrichment media	
selenite broth	peptone 0.5%; mannitol 0.4%; disodium hydrogen phosphate 1%; sodium hydrogen selenite ($NaHSeO_3$) 0.4%
Differential media	
MacConkey's agar	MacConkey's broth (containing neutral red) gelled with 1.5-2.0% agar
Transport media	
Stuart's transport medium	salts; agar 0.2-1.0%; sodium thioglycollate; a redox indicator

NB. There is some overlap between the various categories of media; for example, selenite broth could equally well be described as a selective medium.

Some media are useful because they do *not* support the growth of certain species. Such media generally contain substances which are intended to suppress the growth of some species while permitting others to grow; a medium of this type is called a *selective medium*. An example of such a medium is MacConkey's broth (Table 4-1). In this medium bile salts inhibit the growth of non-enteric bacteria while enteric species can grow freely — allowing their separation from an inoculum containing both enteric and non-enteric bacteria. (Enteric bacteria include many members of the family Enterobacteriaceae.)

An *enrichment medium* (not to be confused with an enriched medium — see above) is a liquid medium in which some species can outgrow others. Such a medium is used, for example, when the inoculum is expected to contain only a minute proportion of the required species among over-whelming numbers of unwanted organisms. An enrichment medium may contain substances which inhibit the growth of unwanted species; alternatively, it may contain substances which encourage the growth of required species without necessarily inhibiting unwanted organisms. For example, selenite broth is a medium which inhibits many types of enteric bacteria (eg. *Eschericia coli*) but which does not inhibit certain species of *Salmonella* — including *S. typhi,* the causal agent of typhoid. Suppose, for example, we need to detect cells of *S. typhi* in a specimen of faeces from a suspected case of typhoid. The specimen may contain very few cells of *S. typhi* so that it may be difficult or impossible to detect them among the vast numbers of non-pathogenic enteric bacteria in the faecal specimen. However, if an inoculum from the specimen is incubated in selenite broth, the proportion of cells of *S. typhi* increases to the point at which they can be detected more readily.

In many cases it is necessary to use a solid medium rather than a liquid medium — for example, to obtain colonies (Chapter 3) of a particular species. Some solid media are simply liquid media to which has been added a gelling agent such as gelatin or *agar*. Agar is a complex polysaccharide obtained from certain seaweeds; it is the most commonly used gelling agent in bacteriological media since (i) it cannot be attacked by the vast majority of bacteria, and (ii) it does not melt at 37°C — a temperature used for the incubation of many types of bacteria. (By contrast, gelatin can be liquefied by some bacteria and is molten at 37°C.) One widely-used agar-based medium is nutrient agar: nutrient broth gelled with 1.5-2.0% agar. Nutrient agar is a general purpose medium which can be used for the culture of a number of species of bacteria; it can also be enriched and/or made selective by the inclusion of appropriate substances. MacConkey's agar (MacConkey's broth gelled with 1.5-2.0% agar) is an example of a *differential medium* — ie. one on which different species may be distinguished by their different forms of growth. On this medium lactose-fermenting enteric bacteria such as *Escherichia coli* form red colonies since they produce acidic products which affect the pH indicator (neutral red) in the medium; enteric species which do not ferment lactose (eg. *Salmonella*) give rise to colourless colonies. Since MacConkey's agar does not support the growth of *non*-enteric bacteria it can be regarded as a selective medium as well as a differential one. Deoxycholate-citrate agar (DCA — Table 4-1) is an agar-based medium which is inhibitory to many non-pathogenic enteric bacteria, but most strains of the enteric pathogens *Salmonella* and *Shigella* form colourless colonies on this medium; any lactose-fermenting bacteria which manage to grow on DCA are distinguished by their red colonies. Blood agar is essentially nutrient agar enriched with blood; it is suitable for the growth of some nutritionally fastidious bacteria and can be used to detect haemolysis (Chapter 11). 'Chocolate agar' is blood agar which has been heated to 70-80°C until it has become chocolate brown in colour; this medium is more suitable than blood agar for the growth of certain fastidious pathogens (eg. *Neisseria gonorrhoeae*).

Some solid media do not contain agar or gelatin. For example, Dorset egg is prepared by the heat-coagulation of a mixture of homogenized hens' eggs and saline; it is used eg. as a *maintenance medium* — ie. a medium used for 'storing' the growth of a particular species.

A *transport medium* is used for the transportation or storage of material from which particular organism(s) are subsequently to be isolated. The main purpose of such a medium is to maintain the viability of the organisms contained within the material; a transport medium need not support growth — indeed growth may be disadvantageous since waste products formed may adversely affect the survival of the organisms. One such medium, Stuart's transport medium (Table 4-1), is suitable eg. for a range of anaerobic bacteria and for 'delicate' organisms such as *Neisseria gonorrhoeae*.

Many media contain substances (eg. peptone, tap water) whose *exact* composition is usually unknown. Sometimes the bacteriologist needs to use a medium in which all the constituents — including those in trace amounts — are quantitatively known; such a medium is called a *defined medium* and is made up from known quantities of pure substances — eg. inorganic salts, glucose, amino acids etc. in distilled or de-ionized water. A defined medium would be used eg. when determining the nutritional requirements of a given species of bacterium.

The preparation of media These days most media can be obtained commercially in a dehydrated, powdered form. Such media are simply dissolved in the appropriate volume of water, sterilized, and dispensed to suitable sterile containers; alternatively, a medium may be dispensed to clean containers prior to sterilization. In the case of agar-based media, such as nutrient agar, the powdered medium is mixed with water and steamed to dissolve the agar. The whole is then sterilized (in an autoclave — Chapter 10) and allowed to cool to about 45°C, a temperature at which the agar remains molten. To prepare a *plate,* some 15-20ml of the molten agar medium is poured into a sterile petri dish which, with the lid on, is left undisturbed until the agar sets. (Blood agar plates are made by mixing molten nutrient agar at about 45-50°C with 5-10% by volume of citrated blood before pouring the plates.) Before use a freshly-made plate must be *dried* — ie. excess surface moisture must be allowed to evaporate; this is achieved by leaving the plate — with the lid partly off — in a 37°C incubator for 10-20 minutes. To prepare a nutrient agar *slope* or *slant* (Fig. 4-1) the molten agar medium is allowed to set in a sterile bottle or test-tube inclined at an angle to the horizontal.

Some types of medium cannot be autoclaved because one or more of their constituents are heat-labile. Media in this category include eg. DCA (which is steamed rather than autoclaved) and media containing glucose or other heat-labile sugars; in preparing the latter type of

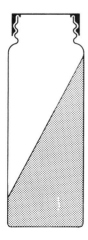

Fig. 4-1 A slope (also called a slant). The medium (shown stippled) has a large surface area available for inoculation; the thickest part is known as the *butt.* Slopes are commonly made of agar-based or gelatin-based media and prepared in small screw-cap bottles; they are used eg. for the storage of a purified strain of bacteria. The surface of a sterile slope is inoculated from a pure culture of the bacterial strain, and the slope is incubated at a suitable temperature to allow growth; it can then be stored in the refrigerator at 4-6°C until required eg. as a source of inoculum.

medium the sugar solution is sterilized separately by membrane filtration (Chapter 10) before being added to the rest of the (autoclaved)
medium.

Aseptic technique

In many bacteriological procedures it is necessary to protect
instruments, sterile media, pure cultures etc from contamination by
organisms that are constantly present in the environment. With this in
mind the bacteriologist takes certain precautions that are collectively
known as an *aseptic technique*. Thus, all tools, vessels, media etc are
sterilized before use, and subsequent contamination — eg. by contact
with non-sterile objects (such as fingers) — must be avoided; sterile
vessels are kept closed except when material is introduced or removed.
When opening a vessel — eg. a sterile vessel or one containing a
medium or culture — the mouth of the vessel is passed briefly through
the bunsen flame in order to kill any extraneous organisms which may
otherwise fall into the vessel and contaminate its contents. This
technique is known as *flaming* and is commonly used, for example,
whenever an inoculum is withdrawn from a culture, or when a sterile
medium is inoculated; flaming is repeated before the vessel is closed.
(Obviously, flaming cannot be used if the vessel in question is a petri
dish, or if the contents of the vessel may catch fire!)

Risks of contamination in the bacteriology laboratory may further
be reduced by treating the benches etc with disinfectants, and by
filtering the air to remove cells and spores of bacteria and fungi etc.
Sometimes bacteriological work is carried out in a cabinet into which
sterile (filtered) air is constantly pumped, and within which instruments and media are manipulated by passing the hands through two
openings in the front of the cabinet; such a procedure protects cultures
etc from air-borne contamination. In another type of cabinet — of
similar appearance — air is drawn into the cabinet and is filtered
before being discharged to the environment; this type of cabinet is
designed to protect the operator against aerosols.

The tools of the bacteriologist

In most cases bacteria can be manipulated with one of the instruments
shown in Fig. 4-2. A loop or straight wire is sterilized immediately
before use by flaming — the wire portion of the instrument being
heated to red heat in a bunsen flame; the wire is then allowed to cool

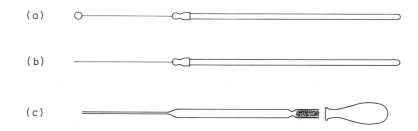

Fig. 4-2 Some basic bacteriological tools. (a) A bacteriological loop. The wire loop may be of platinum, nickel-steel, or nichrome, and is held in a metal handle of about 10-12cm in length. (b) A straight wire: the metal handle carries a straight piece of platinum, nickel-steel or nichrome wire about 5-8cm in length. (c) A pasteur pipette: an open-ended glass tube, the narrow end of which has an internal diameter of about 1mm or less; the wider end is plugged with cotton wool before sterilization. A rubber bulb (teat) is fitted immediately before use.

without touching any non-sterile object. If a flamed and cooled loop is dipped into a suspension of bacteria and withdrawn (using the aseptic technique described above), the loop of wire retains a small circular film of liquid containing a number of bacterial cells, and this can be used as an inoculum. The number of cells withdrawn from the suspension can be varied by using a loop of larger or smaller diameter, or by diluting the suspension; in general, the volume of liquid carried by a loop is about 0.01-0.005ml. Smaller amounts of liquid can be manipulated with the straight wire since this instrument picks up only the minute volume of suspension which adheres to the wire surface. The loop and straight wire can also be used for picking up small quantities of solid material — eg. small amounts of growth from a bacterial colony — simply by bringing the wire loop, or the tip of the straight wire, into contact with the material; the *amount* of material which adheres to the wire will be unknown, but usually this is not important. Liquid or solid inocula carried by a loop or straight wire can be used to inoculate either a liquid or a solid medium; methods of inoculation are considered below. Both the loop and straight wire must always be sterilized by flaming immediately after use so that they do not contaminate the bench and environment. (Note that when an instrument carrying liquid or solid is flamed, spattering may occur and give rise to an aerosol; in some laboratories flaming is carried out with a special bunsen burner which is fitted with a tubular hood — or flaming may be carried out in a special cabinet.)

Larger volumes of liquid may be handled by means of pasteur pipettes or graduated pipettes. Pipettes used in bacteriology are usually plugged with cotton wool (Fig. 4-2), before being sterilized, in order to avoid contamination from the bulb or other mechanical pipetting device. Pipettes are usually sterilized (in batches) inside metal canisters or in thick paper packets; when a pipette is removed from the container, only the wide (plugged) end should be held to avoid contamination of the rest of the pipette. Pasteur pipettes are commonly used once only and are then discarded into a jar of a disinfectant such as lysol or sudol. Graduated pipettes which have been contaminated with bacteria are immersed in a suitable disinfectant until they are safe to handle, when they can be washed up and re-used.

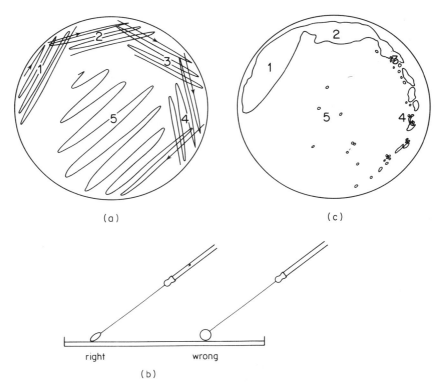

Fig. 4-3 Streaking: inoculating a plate to obtain individual colonies. A loop carrying the inoculum is drawn back and forth (ie. streaked) across a peripheral region of the plate — following the path shown at *1* in diagram (a); when used for streaking, the loop makes contact with the medium as shown in diagram (b). The loop is then flamed, allowed to cool, and streaked across the medium as shown at *2*. Streakings *3*, *4* and *5* are similarly made, the loop being flamed and cooled between each streaking — and after the last. On incubation, areas of the plate inoculated with large numbers of cells give rise to confluent growth — as at *1*, *2* and *3* in diagram (c); well-separated cells give rise to individual colonies — as at *4* and *5*.

Methods of inoculation

Inoculating a liquid medium To inoculate a liquid medium with a *liquid* inoculum, the loop (or straight wire) carrying the inoculum is simply dipped into the liquid medium and withdrawn. With a *solid* inoculum, the loop or straight wire may be rubbed lightly against the inside of the vessel containing the medium; this ensures that at least some of the inoculum is left behind when the instrument is withdrawn.

Inoculating a solid medium Solid media may be inoculated in a variety of ways — particular methods being used for particular purposes. The method of *streaking* (Fig. 4-3) is used when individual, well-separated colonies are required and the (liquid or solid) inoculum is known to contain a large number of cells. In this method the inoculum is progressively 'thinned out' in such a way that individual, well-separated cells are deposited on at least some areas of the plate — usually at 3, 4 or 5 (Fig. 4-3); on incubation each well-separated cell gives rise to an individual colony. In *stab inoculation* a solid medium — eg. the butt of a slope — is inoculated with a straight wire by plunging the wire vertically into the medium; the inoculum (at the tip of the wire) is thus distributed along the length of the stab. Stab inoculation is used eg. for inoculating the deep, anaerobic parts of a medium. A *spread plate* is prepared by spreading a small volume of liquid inoculum (eg. 0.05-0.10ml) over the surface of a solid medium by means of a sterile L-shaped glass rod (a 'spreader'). A *flood plate* is prepared by flooding the surface of a solid medium with a liquid inoculum and withdrawing excess inoculum with a sterile pasteur pipette. If the inoculum contains a sufficient number of cells, incubation of either a spread plate or a flood plate will give rise to a *lawn plate*: ie. a plate in which the agar surface is covered with a layer of confluent growth (Chapter 3). A lawn plate may also be prepared by inoculating a solid medium with a *swab* carrying a liquid or solid inoculum — a swab being a small ball of cotton wool attached to a wooden or wire handle.

The isolation of a species: an application of simple techniques

Some of the basic techniques used in bacteriology can be illustrated by following through a common procedure such as the *isolation* (ie. separation) of one particular species of bacterium from a mixture of organisms. The following account gives a stepwise description of the way in which *Escherichia coli* (Chapter 12) can be isolated from a sample of sewage and obtained in pure culture.

Using a sterile loop — and an aseptic technique — a loopful of sewage is taken from the sample and streaked onto a plate of MacConkey's agar. The plate is incubated for 18-24 hours at 37°C. (Plates are incubated upside-down; if incubated the right way up, water vapour from the medium may condense on the inside of the lid and drop onto the surface of the medium.) During incubation well-separated cells of any species capable of growing on the medium will each give rise to an individual colony (Chapter 3). On MacConkey's agar *E. coli* forms round, red colonies of about 2-3mm in diameter — but not all colonies of this type will necessarily be those of *E. coli*. The next step is to choose several such colonies for further examination; since *E. coli* is common in sewage at least one of the selected colonies is likely to be that of *E. coli*. Before identification can be attempted it is necessary to be sure that each of the selected colonies consists of the cells of only one species; there is a possibility that a particular colony may contain two different species — for example if, during streaking, two cells had been deposited at the same point on the surface of the medium. Purification is achieved by *subculture* — a process in which cells from an existing culture are transferred to a fresh, sterile medium. To subculture a given colony, the *surface* of the colony is lightly touched with a sterile loop so that a small (often invisible) quantity of growth adheres to the loop; this growth is then streaked onto a plate of sterile MacConkey's agar. (Some species of bacteria form minute colonies, and in such cases it is often easier to subculture by touching the surface of the colony with the tip of a straight wire; the inoculum is then carried on the straight wire to a fresh medium where it is streaked by means of a sterile loop.) Each plate — inoculated from a single colony — is then incubated as before. If each of the chosen red colonies was that of a single species (and if the subculturing was properly carried out) we should now have several pure cultures — at least one of which is likely to be that of *E. coli*. (To increase the chances that a given culture is pure it may be subcultured again.) Once a pure culture has been obtained it can be subjected to the identification procedures outlined in Chapter 11 to establish whether or not any of the selected colonies were those of *E. coli*. The pure culture can be stored eg. on a nutrient agar slope as explained in Fig. 4-1.

Incubation under anaerobic conditions

When growing anaerobic bacteria, incubation must obviously be carried out under anaerobic conditions. This can be achieved by using an *anaerobic jar* (sometimes called a McIntosh and Fildes' jar): a strong cylindrical metal chamber with a flat, circular, gas-tight lid. The jar is loaded with a vertical stack of plates (the right way up — ie. lid side up) and the lid of the jar is secured by means of a screw clamp. The jar is then evacuated, by a suction pump, via one of two valves in the lid; this valve is then closed. (If the plates had been put into the jar upside-down, the layer of agar may be sucked from the base of the petri dish by the vacuum.) A rubber bladder, filled with hydrogen, is then attached to the other valve which is opened to permit the entry of gas under atmospheric pressure; this valve is then closed. The cycle of evacuation and re-filling may be carried out several times in succession. Attached to the inside of the lid is a gauze envelope containing a catalyst (eg. palladium-coated pellets of alumina) which promotes chemical combination between hydrogen and the last traces of oxygen. The anaerobic jar is placed inside an incubator for an appropriate period of time.

A more modern form of anaerobic jar consists of a stout cylindrical vessel of strong transparent plastic with a flat, gas-tight lid. The jar is loaded with plates; water is then added to a small packet of chemicals which is dropped into the jar immediately before the latter is closed by means of a screw clamp. The chemicals liberate hydrogen gas which combines with oxygen under the influence of a catalyst. Since in this type of anaerobic jar there is no vacuum, the plates can be inserted upside-down (ie. lid-side down); this avoids the problem of water of condensation dropping from the inside of the lid onto the agar surface.

Most anaerobic jars contain a *redox indicator* which indicates the state of anaerobiosis in the jar. In metal jars the indicator is placed in a small glass side-arm, while in plastic jars an indicator-soaked pad is usually visible through the wall of the jar.

Counting bacteria

The total number of (living and dead) cells in a sample is called the *total cell count,* while the number of living cells is termed the *viable cell count.* Counts in liquid samples are usually given as the number of cells per millilitre (or per 100ml).

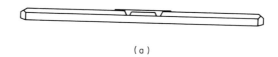

(a)

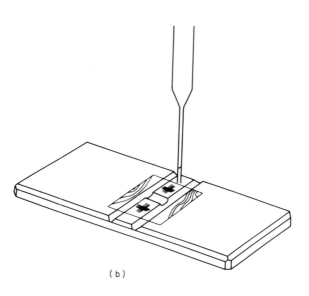

(b)

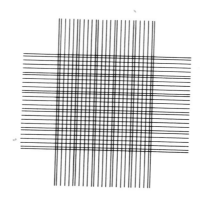

(c)

Fig. 4-4 A typical counting chamber (haemocytometer). The instrument — seen from one side at (a) — consists of a rectangular glass block in which the central plateau lies precisely 0.1mm below the level of the shoulders on either side. The central plateau is separated from each shoulder by a trough, and is itself divided into two parts by a shallow trough (seen at (b)). On the surface of each part of the central plateau is an etched grid (c) consisting of a square which is divided into 400 small squares, each 1/400mm^2. A thin glass cover slip is positioned as shown at (b) and pressed firmly onto the shoulders of the chamber; to achieve proper contact it is necessary, while pressing, to move the cover slip (slightly) against the surface of the shoulders. Proper contact is indicated by the appearance of a pattern of coloured lines (Newton's rings), shown at (b).

Using the chamber A small volume of a bacterial suspension is picked up in a pasteur pipette by capillary attraction; the thread of liquid in the pipette should not be more than 10mm. The pipette is then placed as shown in (b), ie. with the opening of the pipette in contact with the central plateau, and the side of the pipette against the cover slip. With the pipette in this position, liquid is automatically drawn by capillary attraction into the space bounded by the cover slip and one-half of the central plateau; the liquid should not overflow into the trough. (It is sometimes necessary to tap the pipette, *lightly*, against the central plateau to encourage the liquid to enter the chamber.) A second sample can be examined, if required, in the other half of the counting chamber. The chamber is left for 30 minutes to allow the cells to settle, and counting is then carried out under a high power of the microscope — which is focused on the grid of the chamber. Since the volume between grid and cover slip is accurately known, the count of cells per unit volume can be calculated.

A worked example Each small square in the grid is 1/400mm^2. As the distance between grid and cover slip is 1/10mm, the volume of liquid over each small square is 1/4000mm^3 — ie. 1/4,000,000cm^3 (since 1cm^3 = 1000mm^3); since 1cm^3 = 1 millilitre (ml), *the volume of liquid over each small square is 1/4,000,000ml.*

Suppose, for example, that on scanning all 400 small squares 500 cells were counted, ie. an average of 500 ÷ 400 (= 1.25) cells per small square. Since 1.25 cells occur in 1/4,000,000ml, the sample must contain 1.25 × 4,000,000 cells per ml, ie. 5,000,000 cells per ml (5 × 10^6 cells ml^{-1}).

If there are too many cells to count in 400 squares, it is possible to count the cells in, say, 9 blocks of 16 small squares — for example, the first, third and fifth blocks of the first, third and fifth rows, ie. a total of 9 × 16 = 144 small squares. Suppose that 317 cells are counted in 144 small squares. This gives an average of 317 ÷ 144 (= 2.2) cells per small square, ie. a count of 2.2 × 4,000,000 = 8.8 × 10^6 cells ml^{-1}.

If the sample has been diluted before examination in the counting chamber, the count obtained must be multiplied by the dilution factor; eg. if diluted 1 in 10, the count is multiplied by 10.

(N.B. The counting chamber discussed here is a Thoma chamber. Sometimes a *Helber* chamber is used for counting bacteria; this chamber differs from the Thoma chamber in that the depth is 0.02mm instead of 0.1mm — but otherwise the principle is identical.)

The total cell count in a liquid sample (eg. a suspension of cells) can be estimated by comparing the *turbidity* of the sample with that of each of a set of tubes (Brown's tubes) which contain suspensions of barium sulphate in increasing concentration; the tubes range from transparent (tube number 1), through translucent, to turbid and opaque (tube number 10). For a given species of bacterium the turbidity of a particular tube corresponds to the turbidity of a suspension of bacteria of known concentration. When using Brown's tubes the sample should be examined in a tube of size and thickness equivalent to those containing the standard suspensions; the turbidity of the sample is then matched, visually, with that of a particular tube, and the concentration of the sample can be read off from a table supplied with the tubes. Another method of estimating the total cell count is to count the individual cells in a small volume of the sample (or in a small volume of the diluted sample) which is introduced into a *counting chamber* (Fig. 4-4) and examined under the microscope.

Most methods of estimating the viable cell count involve the inoculation of a solid medium with the sample or diluted sample. After incubation, the number of cells in the inoculum can be estimated from the number of colonies which develop on or within the medium; it is always assumed that each colony has arisen from a single cell. (The number of cells which actually give rise to colonies is determined at least partly by the type of medium used and /or by the conditions of incubation.) In the *spread plate* (or *surface spread*) method an inoculum of about 0.05-0.10ml of the sample, or diluted sample, is spread over the surface of a sterile plate as described earlier; the plate is dried, incubated, and the viable cell count is calculated from (i) the number of colonies, (ii) the volume of inoculum used, and (iii) the degree (if any) to which the sample was diluted in order to obtain the inoculum. (If a sample is suspected of containing a high concentration of cells — eg. 10^6 cells per millilitre — the sample can be diluted in 10-fold steps and each dilution inoculated onto a separate plate; at least one dilution will give a countable number of colonies.) In the *pour plate* method a known volume of the sample (or a dilution of it) is mixed with molten agar-based medium (at about 45°C) which is then allowed to set in a petri dish; on incubation, colonies develop within — as well as on — the medium, and the viable cell count is calculated as in the spread plate method above. Yet another method of estimating the viable cell count is known as *Miles and Misra's method* which is described in Fig. 4-5.

Samples which may contain small numbers of bacteria (eg. stream

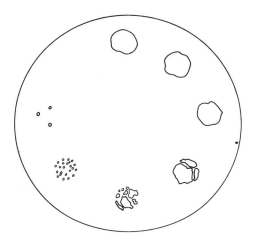

Fig. 4-5 Miles and Misra's method (the drop method) for estimating viable cell counts. This method may be used when a liquid sample is likely to contain a large number of viable cells. The sample is first diluted — usually in 10-fold steps, i.e 1/10, 1/100 ... etc. One drop (of known volume) from each dilution is then placed at a separate, recorded position on the surface of a dried plate of suitable medium. The drops are allowed to dry, and the plate is incubated until visible growth develops on the small circular areas formerly occupied by the drops. Drops which contained large numbers of viable cells will give rise to circular areas of confluent growth. Any drop which contained less than about 15-20 viable cells will give rise to a small, countable number of colonies; by assuming that each colony arose from a single viable cell, the viable count can be estimated from (i) the number of colonies, (ii) the volume of the drop, and (iii) the dilution factor.

NB. Drops from each of the dilutions can be delivered to the plate by means of the same pasteur pipette *if* the drops are taken first from the highest dilution, then from the next highest dilution, and so on. If this is done the pipette can be *calibrated* after use, ie. the volume of the drops delivered by the pipette can be estimated. This is done by drawing into the pipette a measured volume (eg. 1ml) of water, and counting the number of drops formed when this volume is discharged; if 1ml is discharged in, say, 30 drops, each drop has a volume of approximately 1/30ml.

water, river water) can be filtered through a sterile membrane of pore size about 0.3-0.45μm which retains the cells; volumes up to 100ml or more may be filtered. The membrane is then placed, cell side uppermost, onto the surface of a solid nutrient medium, and incubated; nutrients diffuse through the pores of the membrane, and individual colonies develop from those cells capable of growth under such conditions. The concentration of viable cells in the sample can then be estimated from (i) the number of colonies formed on the membrane, and (ii) the volume of the sample filtered.

5 Bacterial metabolism

Metabolism can be thought of as the 'mechanism of life'. It involves the breaking down, building up, and interconversion of molecules within the cell by a whole range of chemical reactions — each reaction being catalysed by a specific protein catalyst (an *enzyme*). A sequence of such reactions — in which one compound is converted to another (or others) — is called a *metabolic pathway*. Some metabolic pathways are common to both bacteria and eukaryotic cells, but there are also many pathways which occur only in bacteria.

Many metabolic reactions require an input of energy. The cell also needs energy for other purposes — for example, movement (in motile species), or the uptake of nutrients against a concentration gradient. All of this energy must be obtained by the cell from its environment, and the way in which this is achieved forms an important part of metabolism; we shall therefore discuss energy metabolism at some length in the present chapter.

The majority of bacteria derive their energy from chemicals obtained from their environment, and such bacteria are called *chemotrophs;* there are also some bacteria which can harness the energy of sunlight, and these are called *phototrophs*. In neither case can the environmental source of energy be used in its original form: light, for example, cannot *directly* fuel energy-requiring metabolic reactions or cell motility. Energy from an environmental source must first be converted by the cell into an intermediate form or 'currency'; energy in currency form can be used by the cell to drive its various energy-requiring processes. A major part of energy metabolism is concerned with the conversion of energy from the environment into currency form.

A compound which plays a key role as an energy currency molecule is adenosine 5'-triphosphate, better known as ATP: Fig. 5-1(a). ATP

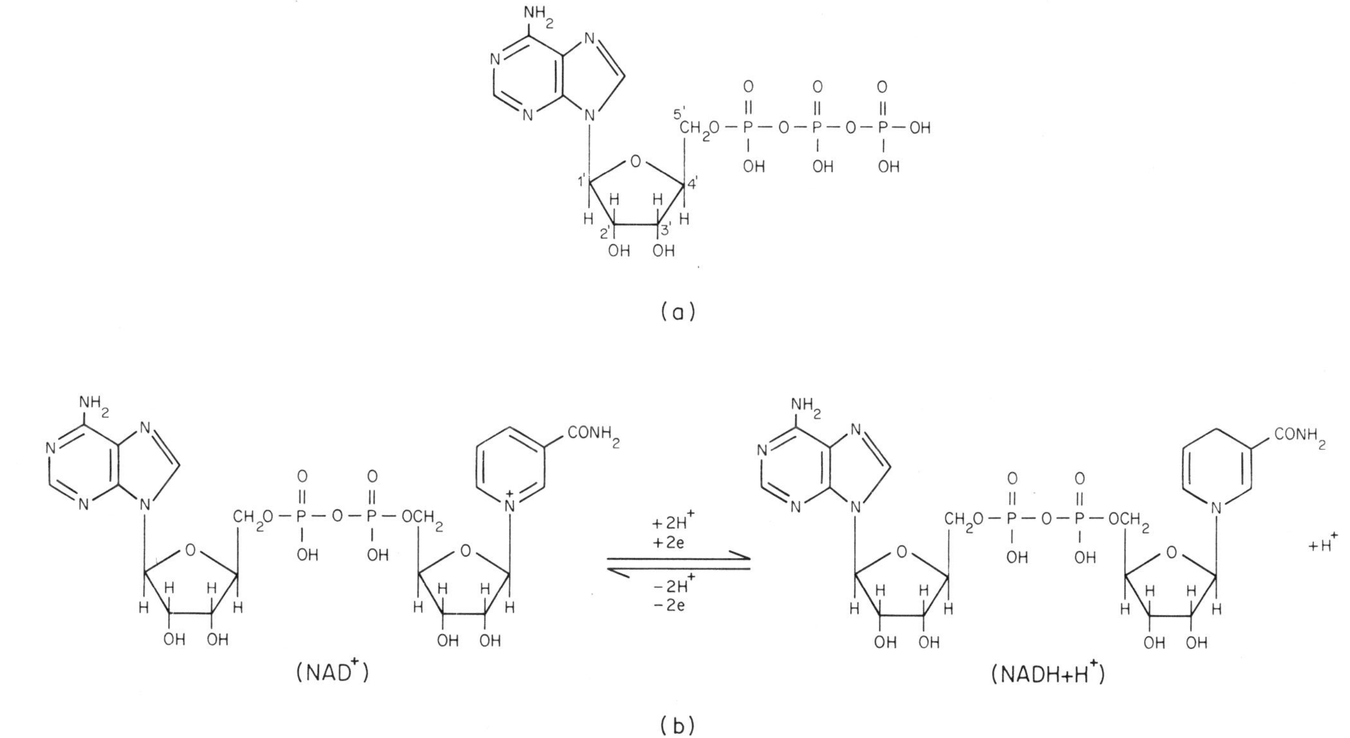

Fig. 5-1 Some currency molecules. (a) Adenosine 5′-triphosphate, ATP. Adenosine 5′-diphosphate (ADP) has two phosphate groups instead of three, while adenosine 5′-monophosphate (AMP) has only one. (b) Nicotinamide adenine dinucleotide (NAD), showing the oxidized and reduced forms. The oxidized form is correctly written NAD⁺ (although it may be written NAD), while the reduced form is correctly written as NADH + H⁺ (often abbreviated to NADH). (e = electron).

releases energy in a form useful to the cell when its terminal phosphate bond is broken to form adenosine 5′-diphosphate, ADP. Thus as molecules of ATP are used to supply the energy needs of the cell, molecules of ADP are formed; environmental energy must therefore be harnessed in a way that allows the re-synthesis of ATP from ADP. The principal processes by which this is achieved will be considered in the following pages.

There is another type of currency molecule in which energy is carried in the form of 'reducing power'. A molecule of this type can accept electrons (and hence energy) from eg. an intermediate in a metabolic pathway; the reduced currency molecule can then either transfer its electrons to another substrate, or — in many bacteria — it can be reoxidized in such a way that its energy can be used for the generation of ATP (see later). A common currency molecule of this type is *nicotinamide adenine dinucleotide,* NAD: Fig. 5-1(b); many examples of the role of NAD will be found later in the chapter. Other carriers of 'reducing power' are *NAD phosphate* (NADP) and flavin-containing compounds such as *flavin adenine dinucleotide* (FAD).

METABOLISM IN CHEMOTROPHIC BACTERIA

Chemotrophs can be divided into two groups: the *chemoorganotrophs* — which obtain energy from the metabolism of organic compounds, and the *chemolithotrophs* — which derive energy from the oxidation of inorganic compounds or elements. The chemolithotrophic bacteria are unique in the biological world: no other group of organisms can carry out this type of metabolism.

The terms 'chemoorganotroph' and 'chemolithotroph' refer only to the way in which *energy* is obtained; they say nothing about the source of *carbon* which the organism uses to make cell components, enzymes etc. Bacteria which can use carbon dioxide as the principal or sole source of carbon are called *autotrophs,* while those which obtain the bulk of their carbon from organic compounds are called *heterotrophs.* A chemolithotroph which uses carbon dioxide as its main or sole source of carbon is called an autotrophic chemolithotroph, whereas a chemolithotroph which obtains its carbon chiefly from organic compounds is referred to as an heterotrophic chemolithotroph. Chemoorganotrophic bacteria are usually or always heterotrophic — obtaining both carbon and energy from organic nutrients — although most can also assimilate some carbon dioxide.

Metabolism in chemoorganotrophs

The chemoorganotrophic bacteria, taken together, can use a vast range of organic compounds as nutrients — although any particular species may be able to use only a very limited number of compounds. Usually, nutrients are taken into the cell from the surrounding medium by means of specific transport mechanisms in the cell membrane (Chapter 2). Once inside the cell the fate of a nutrient depends on the nature of the nutrient, on the species of bacterium and its requirements, and on environmental conditions. Nutrients which are used for the production of energy are metabolized by one of two basic types of energy-yielding process: *fermentation* and *respiration*.

Fermentation Fermentation is a type of energy-yielding metabolism which typically occurs in the absence of oxygen. An organic substrate (eg. glucose) is metabolized via a particular pathway, and energy released by certain (exergonic) reactions in that pathway is used for the synthesis of ATP. When energy from a chemical reaction is the driving force for the synthesis of ATP from ADP the process is referred to as *substrate-level phosphorylation*. It occurs, for example, in a reaction of the type:

$$X\text{-}P + ADP \quad \rightarrow \quad X\text{-}H + ATP$$

where X-P is an energy-rich organic phosphate which is formed as an intermediate in the pathway. Substrate-level phosphorylation is the only way in which energy may be obtained by fermentation.

The fermentation of a substrate can never involve the net oxidation of that substrate. Net oxidation would require the use of an electron 'sink' — ie. an oxidizing agent, obtained from the environment, which could accept electrons removed from the substrate. No such oxidizing agent is involved in fermentation (compare with respiration: see later). If any intermediate in a fermentation pathway undergoes oxidation, this oxidation must necessarily be balanced by equivalent reduction of

Fig. 5-2 Fermentation: a generalized scheme to illustrate the overall oxidation-reduction balance in a fermentation pathway. (e = electron).

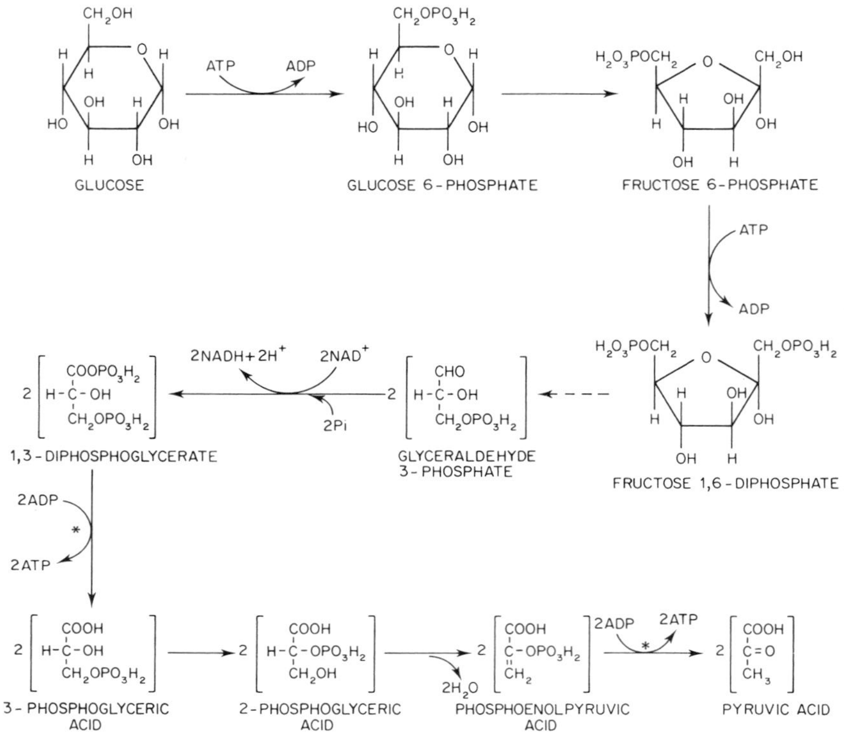

Fig. 5-3 The Embden-Meyerhof pathway for the breakdown of glucose to pyruvic acid. The broken arrow (← – – –) indicates that more than one reaction is involved. Substrate-level phosphorylations are marked by an asterisk (*). (Pi = inorganic phosphate, PO_4^{3-}).

other intermediate(s) in the pathway (Fig. 5-2); this means that the products of fermentation — taken together — must have the same oxidation state as the original substrate.

These somewhat abstract principles are much more easily understood when specific examples of fermentation are considered. Many bacteria can carry out fermentation using glucose as a substrate, and the process often begins with a sequence of reactions which is referred to as the *Embden-Meyerhof pathway* or *glycolysis* (Fig. 5-3). This pathway may be summarized by the equation:

Glucose + 2ATP + 4ADP + 2Pi + 2NAD$^+$ → 2Pyruvic acid + 2ADP + 4ATP + 2NADH + 2H$^+$ + 2H$_2$0

Thus, each molecule of glucose metabolized gives a *net* yield of 2ATP (formed by substrate-level phosphorylation). The repetition of this sequence of reactions for many glucose molecules requires fresh supplies of ADP and NAD. The ADP is regenerated from ATP when the latter is used to supply the cell with energy. However, the way in which NAD is regenerated from NADH may differ from one species to another. In the simplest case, NADH donates its electrons to pyruvic acid — ie. the oxidation of NADH to NAD is coupled to the reduction of pyruvic acid to lactic acid:

$$CH_3.CO.COOH + NADH + H^+ \rightarrow CH_3.CHOH.COOH + NAD^+$$

Pyruvic acid   Lactic acid

This reaction completes one possible fermentation pathway. Note that the molecular formula of lactic acid is $C_3H_6O_3$: ie. it has the same oxidation state as glucose ($C_6H_{12}O_6$); no net oxidation or reduction has occured. However, the pathway from glucose to pyruvic acid involves the oxidation of glyceraldehyde 3-phosphate, but this oxidation is later balanced by the reduction of pyruvic acid to lactic acid. The complete pathway from glucose to lactic acid is called *homolactic fermentation* and occurs in some species of *Streptococcus* and *Lactobacillus.* Lactic acid is a waste product and is released by the cells — causing the medium to become increasingly acidic. Some bacteria which produce lactic acid in this way are used in the manufacture of dairy products such as yoghurt; milk is inoculated with species of *Lactobacillus* and *Streptococcus* which ferment the milk sugar to produce lactic acid. The lactic acid acts both as a preservative (by lowering the pH) and as a flavour component in the resulting product.

Two other fermentation pathways which involve the metabolism of glucose by the Embden-Meyerhof pathway are the *mixed acid fermentation* and the *butanediol fermentation.* These pathways both differ from the homolactic fermentation in that several end-products are formed from pyruvic acid in relative proportions which depend on environmental conditions. The mixed acid fermentation (Fig. 5-4) occurs in eg. *Escherichia coli* and species of *Proteus, Salmonella, Shigella* and *Yersinia,* while the butanediol fermentation (Fig. 5-5) occurs eg. in species of *Klebsiella* and *Serratia.* Although these two fermentation pathways are more complicated than the homolactic fermentation, they nevertheless conform to the same basic principles of fermentation. Thus, in each fermentation pathway, although the

52

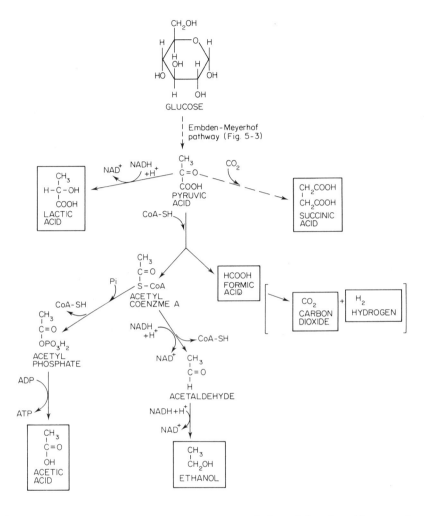

Fig. 5-4 The mixed acid fermentation. The splitting of formic acid to produce carbon dioxide and hydrogen occurs only in some species and strains; this reaction is responsible for gas production from glucose metabolism in these bacteria. (CoA-SH = coenzyme A; Pi = inorganic phosphate, PO_4^{3-}).

relative proportions of the end-products may vary, the formation of products more oxidized than glucose (such as formic acid: CH_2O_2) must always be balanced by the formation of products more reduced than glucose (such as ethanol: C_2H_6O). Again, as in all fermentations, ATP is formed only by substrate-level phosphorylation — as, for example, when acetic acid is formed from acetylphosphate (Fig. 5-4).

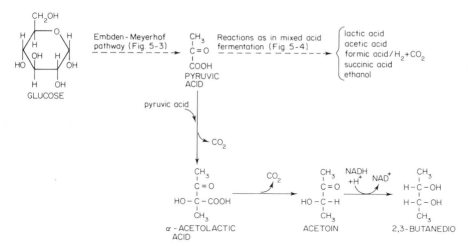

Fig. 5-5   The butanediol fermentation.

Respiration In chemoorganotrophs respiration is an energy-yielding process in which an organic substrate is oxidized by means of an oxidizing agent obtained from the environment. Glucose, for example, is commonly oxidized completely to carbon dioxide and water. Oxidation occurs in a number of stages in which electrons are transferred to currency molecules such as NAD; from the reduced currency molecules these electrons are transferred, indirectly, to the oxidizing agent (the *terminal electron acceptor*) — see Fig. 5-6. The terminal electron acceptor most commonly used is oxygen; hence, respiration is usually — though not always — an aerobic process. The complete oxidation of an organic substrate by respiration allows a cell to obtain far more energy than it could obtain from the same substrate by fermentation. This can be illustrated by considering the respiration of glucose and comparing the energy yield with that of a fermentation pathway.

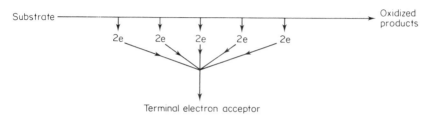

Fig. 5-6   Respiration: a generalized scheme illustrating the role of an external electron acceptor. (Compare with Fig. 5-2).

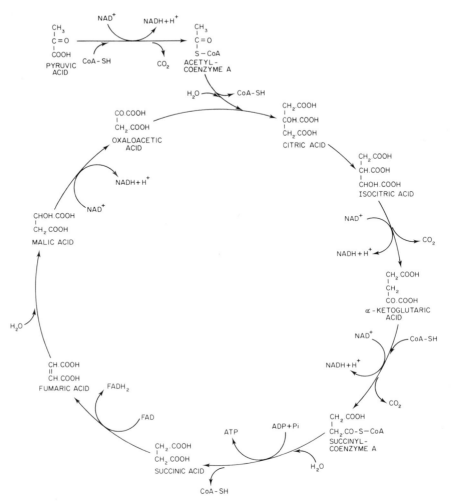

Fig. 5-7 The oxidation of pyruvic acid via the tricarboxylic acid cycle. (CoA-SH = coenzyme A; Pi = inorganic phosphate, PO_4^{3-}).

(a) *Aerobic respiration.* In many bacteria the early reactions in the respiratory metabolism of glucose are the same as those in many types of fermentation: glucose is broken down to pyruvic acid via the Embden-Meyerhof pathway (Fig. 5-3). Beyond pyruvic acid, however, the pathways of fermentation and respiration are quite different. In a typical respiratory pathway, pyruvic acid is converted to acetyl-coenzyme A which is fed into the *tricarboxylic acid cycle* (*Krebs' cycle, citric acid cycle*) — see Fig. 5-7. In one complete turn of this cycle pyruvic acid is, in effect, oxidized to carbon dioxide:

$$CH_3.CO.COOH + 4NAD^+ + FAD + ADP + Pi +$$
$$3H_2O \rightarrow 3CO_2 + 4NADH + 4H^+ + FADH_2 + ATP$$

The cell now has $4NADH$ and $1FADH_2$ formed for each molecule of pyruvic acid oxidized via the tricarboxylic acid cycle, as well as the NADH formed during the Embden-Meyerhof pathway; all of these reduced currency molecules must be re-oxidized to allow respiration to continue. This re-oxidation is achieved by the transfer of electrons from the reduced currency molecules to oxygen via a chain of carrier molecules called the *electron transport chain* or *respiratory chain*. The respiratory chain forms a conducting path along which electrons can flow as a result of sequential reduction/oxidation reactions: Fig. 5-8.

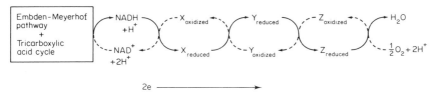

Fig. 5-8 The flow of electrons (e) along an aerobic respiratory chain. X, Y and Z are electron carriers; the solid curved lines indicate the path of electron flow.

The respiratory chain in a bacterium is located in the cell membrane and usually consists of cytochromes (haem-containing proteins), quinones, and other components. However, in bacteria the nature of the respiratory chain can vary from one species to another, and may vary even in one species growing under different conditions.

The respiratory chain provides a means whereby the re-oxidation of NADH etc. can yield energy for use by the cell — ie. the transfer of electrons along a respiratory chain is itself an energy-yielding process. The energy is initially conserved in the form of a 'high energy intermediate' which can, in turn, be used to make ATP from ADP and inorganic phosphate (Fig. 5-9). The formation of ATP in this way is called *oxidative phosphorylation*. The high energy intermediate can also supply energy directly for certain other energy-requiring processes — including, for example, flagellar motility and the transport of ions and some substrates across the cell membrane — see Fig. 5-9.

The precise nature of the high energy intermediate has long been a subject for debate. One theory which is supported by a considerable amount of evidence is the *chemiosmotic theory* developed chiefly by the British biochemist Peter Mitchell. According to this theory the

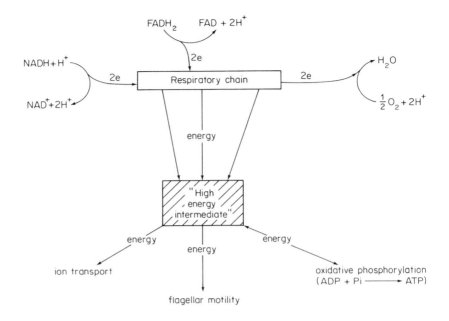

Fig. 5-9 The role of a "high energy intermediate" in the conservation of energy from electron transport. (Note that energy may also be transferred *to* the high energy intermediate from ATP by ATP hydrolysis.)

transfer of electrons along a respiratory chain causes hydrogen ions (protons) from the cytoplasm to be pumped outwards across the cell membrane. Since the cell membrane is ordinarily impermeable to ions this leads to an imbalance of pH and electrical charge across the membrane — the cytoplasm becoming alkaline and electrically negative with respect to the outer surface of the cell membrane. This results in the generation of a force (called the *proton-motive force*) which tends to pull protons back into the cytoplasm. The resulting flow of protons back into the cell is accompanied by the performance of work — such as, for example, flagellar motility, the transport of ions and other substrates across the cell membrane, and the phosphorylation of ADP. The proton-motive force therefore functions as the 'high energy intermediate' mentioned above.

We are now in a position to compare the energy yields of aerobic respiration and fermentation. During respiration the amount of ATP yielded by oxidative phosphorylation may vary with the nature of the respiratory chain; in general, a maximum of 3ATP may be formed for each NADH oxidized, while 2ATP is generally the maximum for each

FADH$_2$ oxidized. For each molecule of *glucose* metabolized by respiration, 2NADH are formed by the Embden-Meyerhof pathway (Fig. 5-3), while 8NADH and 2FADH$_2$ are formed by the tricarboxylic acid cycle (Fig. 5-7). This gives a maximum yield of 34ATP formed by oxidative phosphorylation. In addition to this, 4ATP per glucose molecule are formed by substrate level phosphorylation: 2ATP from the Embden-Meyerhof pathway, and 2ATP (one for each pyruvic acid molecule) from the tricarboxylic acid cycle. Thus, for each molecule of glucose metabolized by aerobic respiration there is a maximum total yield of 38ATP. This compares with, for example, a yield of only 2ATP (formed by substrate-level phosphorylation) per molecule of glucose metabolized by homolactic fermentation.

In a living cell, of course, no pathway or process can occur in isolation but rather forms one part of a complex network of interacting pathways. Thus, the theoretical maximum yield of ATP in respiration may not be reached in the living bacterium owing to interactions between the respiratory pathway and other pathways or processes. For example: (i) not all of the energy generated by electron transport is used for oxidative phosphorylation. As well as driving oxidative phosphorylation, the proton-motive force is also used directly to drive eg. flagellar motility and ion transport. (ii) Some of the NADH generated during glucose oxidation may be used as a source of reducing power for certain biosynthetic reactions; this NADH is therefore not available for oxidation via the respiratory chain. (iii) Intermediates in the Embden-Meyerhof pathway and the tricarboxylic acid cycle may be drawn off to supply the cell with 'building blocks' for biosynthetic pathways. (For example, pyruvic acid is used in the synthesis of the amino acids alanine, valine and leucine.) The withdrawal of such intermediates necessarily involves the sacrifice of the energy which could have been obtained by the complete oxidation of those intermediates.

(b) *Anaerobic respiration.* In some bacteria respiration can occur under anaerobic conditions. In principle anaerobic respiration is similar to aerobic respiration, but in the absence of oxygen some other terminal electron acceptor must be used. Many bacteria can use nitrate as an alternative to oxygen in a process called *nitrate respiration.* Nitrate respiration allows an organism such as *Pseudomonas aeruginosa* — which is capable only of respiratory metabolism — to grow under anaerobic conditions in the presence of nitrate. Certain other facultative anaerobes — including many members of the family Enterobacteriaceae (Chapter 12) — can also carry out nitrate respiration. (In the absence of both oxygen and nitrate members of the Enterobacteriaceae can obtain energy by fermentation.) Nitrate respiration involves many of the components of the aerobic respiratory chain; as in aerobic respiration, electron transport is accompanied by the generation of a proton-motive force, so that ATP can still be formed by oxidative phosphorylation. In general, however, the energy yielded by nitrate respiration is lower than that yielded by aerobic respiration. The transfer of electrons to nitrate results in the formation of nitrite:

$$NO_3^- + 2e + 2H^+ \rightarrow NO_2^- + H_2O$$

Some bacteria can further reduce nitrite to gaseous nitrogen (a process referred to as *denitrification*), while in others the end product is ammonia.

Bacteria of the genera *Desulfovibrio* and *Desulfotomaculum* are obligate anaerobes which carry out only respiratory metabolism using sulphate as the terminal electron acceptor. Sulphate is reduced to sulphide via sulphite:

$$SO_4^{2-} + 2e + 2H^+ \rightarrow SO_3^{2-} + H_2O$$
$$SO_3^{2-} + 6e + 6H^+ \rightarrow S^{2-} + 3H_2O$$

Some of the sulphide may be used to fulfill the cell's sulphur requirement (eg. for the synthesis of sulphur-containing amino acids such as cysteine) while the remainder is lost to the environment as hydrogen sulphide.

Metabolism in chemolithotrophs

Chemolithotrophic bacteria obtain their energy by the oxidation of inorganic substances such as ammonia, nitrite, sulphide, hydrogen, and ferrous iron. Chemolithotrophic metabolism commonly occurs under aerobic conditions and is essentially respiratory in nature; electrons derived from the inorganic substances are fed into a respiratory chain through which they are transferred to an electron acceptor such as oxygen. As in chemoorganotrophic respiration, energy is conserved in the form of a proton-motive force which can be used to drive ATP formation.

Bacteria which can obtain energy by the oxidation of ammonia or nitrite (*nitrification*) play an important role in the nitrogen cycle

(Chapter 8). These are the *nitrifying bacteria* — Gram-negative, obligately aerobic, autotrophic chemolithotrophs found in soil and water. Some nitrifying bacteria (eg. *Nitrosomonas, Nitrosococcus*) oxidize ammonia to nitrite, while others (eg. *Nitrobacter, Nitrococcus*) oxidize nitrite to nitrate.

Members of the genus *Thiobacillus* derive energy from the oxidation of sulphur and inorganic sulphur compounds (eg. sulphide, thiosulphate). The terminal electron acceptor is usually oxygen, although *Thiobacillus denitrificans* can use nitrate as an alternative electron acceptor under anaerobic conditions. The thiobacilli occur in a range of sulphur-containing habitats and play an important part in the sulphur cycle (Chapter 8).

A number of bacteria can obtain energy by the oxidation of hydrogen (H_2). For example, *Alcaligenes eutropha* and *Pseudomonas facilis* can carry out autotrophic chemolithotrophy in the presence of hydrogen (as electron donor), oxygen (as electron acceptor), and carbon dioxide (as carbon source). The family Methanobacteriaceae contains anaerobic bacteria which oxidize hydrogen using carbon dioxide as the terminal electron acceptor; the products include methane:

$$4H_2 + CO_2 \rightarrow CH_4 + 2H_2O$$

Such methane-producing bacteria occur in the anaerobic muds of ponds, in sewage sludge, etc.

METABOLISM IN PHOTOTROPHIC BACTERIA

As discussed earlier, phototrophs obtain their energy from sunlight. Nearly all phototrophic bacteria — and the cyanobacteria — achieve this by *photosynthesis:* a process which involves the trapping of light energy by green pigments called chlorophylls. The exceptional type of phototrophy, which does not involve chlorophylls, occurs in certain bacteria of the genus *Halobacterium*. A phototroph may obtain its carbon from carbon dioxide (a photoautotroph) or organic compounds (a photoheterotroph).

Photosynthesis in bacteria

The photosynthetic bacteria (excluding for the moment the cyanobacteria) fall mainly into three families: the Chlorobiaceae (green sulphur bacteria), the Chromatiaceae (purple sulphur bacteria), and the Rhodospirillaceae (purple non-sulphur bacteria). (There is also a new fourth family, the Chloroflexaceae, which contains gliding, filamentous, photosynthetic bacteria which appear to be related to the green sulphur bacteria.) All of these bacteria are Gram-negative organisms which occur chiefly in anaerobic aquatic environments. Photosynthesis in these bacteria differs in several respects from that in green plants. Perhaps the most obvious difference is that bacterial photosynthesis never involves the evolution of oxygen, and can in fact occur only under strictly anaerobic conditions.

The chlorophylls in photosynthetic bacteria differ slightly in structure from plant chlorophylls and so are referred to as *bacteriochlorophylls*. Bacteriochlorophylls — together with other components of the photosynthetic apparatus — are bound to specialized membrane systems which form vesicles inside the cells. (Bacteria, being prokaryotes, do not contain chloroplasts.) In the green bacteria (eg. *Chlorobium*) the membrane vesicles are attached to — but distinct in structure from — the cell membrane; they are called *chlorobium vesicles*. In both types of purple bacteria the membrane vesicles are continuous with the cell membrane and are called *chromatophores* or *thylakoids*.

The photosynthetic pigments (bacteriochlorophylls and carotenoids) absorb energy from sunlight and transfer it to a special group of bacteriochlorophylls which make up a so-called *reaction centre*. On receiving this energy the reaction centre ejects energized electrons which are accepted by a primary electron acceptor. These electrons may subsequently follow a cyclic pathway — being transferred via a sequence of electron carriers (quinones, cytochromes) back to the reaction centre. In the purple bacteria the cyclic flow of electrons is associated with the generation of a proton-motive force across the chromatophore membranes — protons accumulating within the chromatophore vesicles. In the green bacteria there is some evidence that the proton-motive force is generated across the cell membrane. In either case the proton-motive force can drive the formation of ATP in a process called *photophosphorylation* — a process analogous to the oxidative phosphorylation associated with respiration. In an alternative pathway, electrons may be withdrawn from the cyclic pathway and used to reduce NAD — so providing the cell with reducing power. Electrons used for the reduction of NAD are no longer available for recycling, and so an electron donor is required to restore electrons to the reaction centre. In the green sulphur bacteria (Chlorobiaceae) inorganic substances such as sulphide or elemental sulphur function as electron donors; photosynthetic organisms which use inorganic electron donors are called *photolithotrophs*. The purple non-sulphur bacteria (Rhodospirillaceae) use organic compounds as electron donors and so are called *photoorganotrophs*. The purple sulphur bacteria (Chromatiaceae) can obtain electrons from sulphur and inorganic sulphur compounds, but in addition some species can obtain electrons from organic compounds.

So far we have excluded the cyanobacteria from our discussion. This is because photosynthesis in these organisms is quite different from that in the photosynthetic bacteria. The cyanobacteria, like the photosynthetic bacteria, are prokaryotic and therefore lack chloroplasts — having instead an internal system of membranes (thylakoids) which contain the photosynthetic apparatus. However, in other respects photosynthesis in the cyanobacteria resembles more closely that in green plants. Thus the cyanobacteria are strictly aerobic and carry out photosynthesis with the evolution of oxygen; the oxygen is derived from water which functions as the electron donor for the reduction of NADP — ie. the cyanobacteria are photolithotrophs.

Phototrophy in *Halobacterium*

Bacteria of the genus *Halobacterium* are halophilic organisms which occur in salt lakes, in the Dead Sea, and in salted foods, etc. Under conditions of high oxygen tension these organisms carry out chemoorganotrophic respiration. However, under conditions of low oxygen tension in the presence of light, some strains of *Halobacterium halobium* and *H. salinarium* (formerly *H. cutirubrum*) can carry out a unique type of phototrophic metabolism; regions of the cell membrane become differentiated to form the so-called *purple membrane* which contains a purple protein called *bacteriorhodopsin*. (Bacteriorhodopsin contains retinal and derives its name from rhodopsin — the retinal-containing protein light receptor in the human eye.) On receiving light energy, bacteriorhodopsin pumps protons across the purple membrane — thus providing a novel mechanism for generating a proton-motive force which can be used to drive ATP synthesis etc. This system represents the simplest known mechanism whereby a proton-motive force is generated by a living cell, and it is the only known biological system in which light is used directly as a source of energy without the involvement of chlorophyll.

6 Bacterial genetics: an introduction

What dictates the nature of a cell? In other words, what makes one species of bacterium different from another, and how is this difference maintained from one generation to the next? Topics such as these form the subject matter of genetics — some aspects of which we shall be exploring in this chapter.

Everything that a cell is or does results from 'instructions' contained within the cell in the form of a *genetic code*. In bacteria (as in most other organisms) this code is built into the structure of a large molecule called *deoxyribonucleic acid* — DNA — which constitutes the chromosome of a bacterial cell (see Chapter 2). DNA and its relative, RNA (ribonucleic acid), belong to a category of macromolecules called *nucleic acids*. Since both types of nucleic acid play central roles in genetics we shall consider briefly their structure and synthesis before going on to examine their functions in the cell.

The structure of nucleic acids

Nucleic acids are polymers which are made up of 'building blocks' called *nucleotides*. A nucleotide consists of three components: a 5-carbon sugar, a nitrogen-containing base, and one or more phosphate groups — see Fig. 6-1. Nucleotides which contain the sugar ribose are called *ribonucleotides,* while those which contain the sugar 2'-deoxyribose are called *deoxyribonucleotides*. RNA is a polymer of ribonucleotides, each of which contains one of the four nitrogen bases adenine, guanine, cytosine, and uracil (see Fig. 6-2). DNA is a polymer of deoxyribonucleotides, each of which contains one of the four nitrogen bases adenine, guanine, cytosine, and thymine (Fig. 6-2). The naming of these nucleotides (and their corresponding nucleosides) is summarized in Table 6.1.

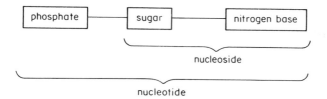

Fig. 6-1 The generalized structure of a nucleotide.

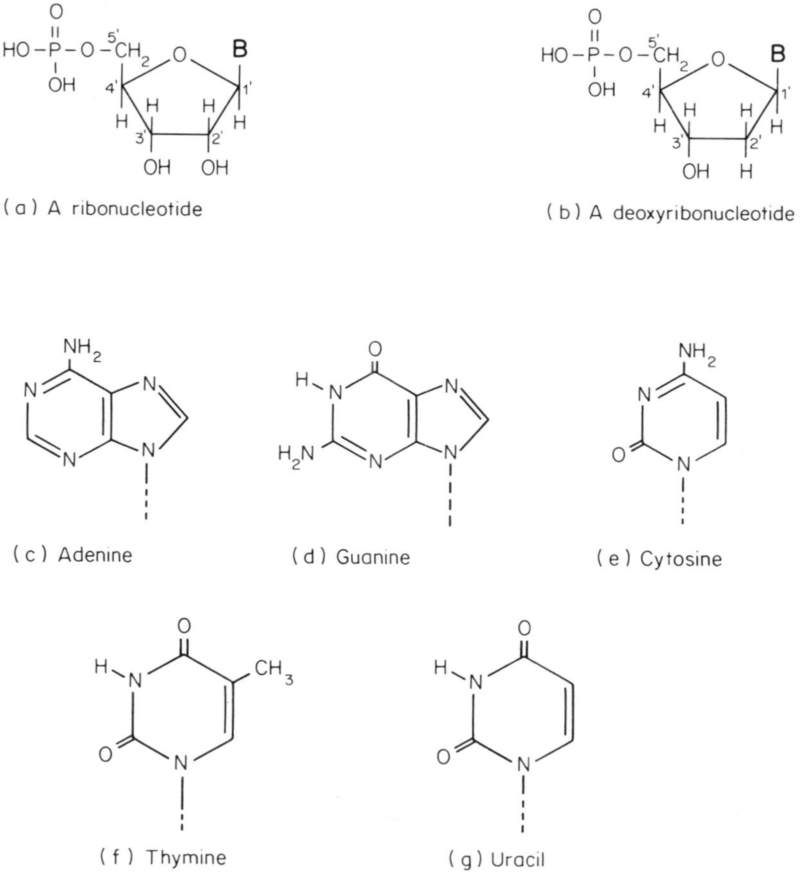

(a) A ribonucleotide

(b) A deoxyribonucleotide

(c) Adenine

(d) Guanine

(e) Cytosine

(f) Thymine

(g) Uracil

Fig. 6-2 The structure of nucleotides in RNA and DNA. (a) A generalized ribonucleotide. In RNA **B** is one of the bases adenine (c), guanine (d), cytosine (e), or uracil (g). (b) A generalized deoxyribonucleotide. In DNA **B** is one of the bases adenine (c), guanine (d), cytosine (e), or thymine (f). (Note: adenine and guanine are substituted *purines*, while cytosine, thymine and uracil are substituted *pyrimidines*.)

Table 6-1 The names of nucleosides and nucleotides containing the bases found in RNA and DNA (Fig. 6-2).

Sugar	Nitrogen base	Nucleoside (base + sugar)	Mononucleotide* (base + sugar + phosphate)
ribose	adenine	adenosine	adenosine 5'-monophosphate (AMP) (or adenylic acid)
ribose	guanine	guanosine	guanosine 5'-monophosphate (GMP) (or guanylic acid)
ribose	cytosine	cytidine	cytidine 5'-monophosphate (CMP) (or cytidylic acid)
ribose	uracil	uridine	uridine 5'-monophosphate (UMP) (or uridylic acid)
deoxyribose	adenine	deoxyadenosine	deoxyadenosine 5'-monophosphate (dAMP) (or deoxyadenylic acid)
deoxyribose	guanine	deoxyguanosine	deoxyguanosine 5'-monophosphate (dGMP) (or deoxyguanylic acid)
deoxyribose	cytosine	deoxycytidine	deoxycytidine 5'-monophosphate (dCMP) (or deoxycytidylic acid)
deoxyribose	thymine	deoxy-thymidine	deoxythymidine 5'-monophosphate (dTMP) (or deoxythymidylic acid)

*A nucleotide is called a *mononucleotide* when it contains one phosphate group, a *dinucleotide* when it contains two phosphate groups, and a *trinucleotide* when it contains three phosphate groups; an example of a trinucleotide is ATP — see Fig. 5-1(a).

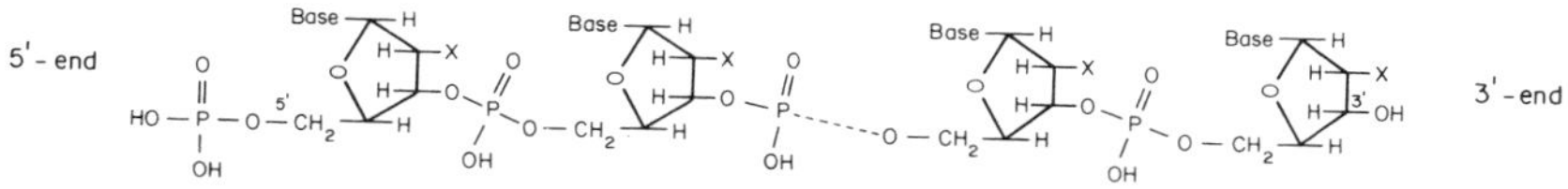

Fig. 6-3 The structure of a single strand of nucleic acid. (X = OH in RNA; X = H in DNA).

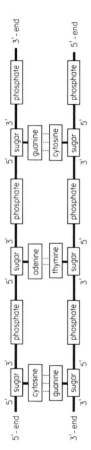

Fig. 6-4 A DNA duplex. Dotted lines represent hydrogen bonds. Note that the two strands are *antiparallel*: ie. a 5′-3′ strand lies alongside a 3′-5′ strand.

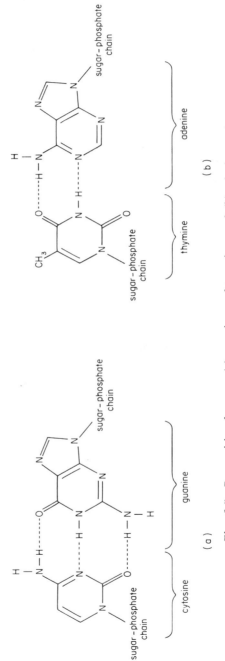

Fig. 6-5 Base-pairing between (a) cytosine and guanine, and (b) adenine and thymine. Dotted lines represent hydrogen bonds.

In a molecule of nucleic acid the nucleotides are linked together to form an unbranched chain or *strand* of alternating sugar and phosphate residues, as shown in Fig. 6-3. In a molecule of DNA there are normally two such strands which are held together by hydrogen bonds between their nitrogen bases; this double-stranded structure is called a DNA *duplex* (Fig. 6-4). The hydrogen bonding between the nitrogen bases is quite specific: adenine pairs only with thymine, while guanine pairs only with cytosine (Fig. 6-5). Each base in such a base pair is said to be *complementary* to its partner; thus adenine is complementary to thymine, while guanine is complementary to cytosine. Hence, in a DNA duplex each strand is complementary to the other in its base sequence (Fig. 6-4) — a feature of great importance, as we shall see later. The DNA duplex is itself twisted into a helix in which one complete turn occurs every ten nucleotides (Fig. 6-6). This is the famous 'double helix' worked out by Watson and Crick in the 1950s. In the bacterial chromosome the helical DNA duplex forms a closed loop — ie. there are no free 3′ or 5′ ends.

Since the two strands in a DNA duplex are complementary to one another it follows that the amount of adenine in the duplex will always be equal to the amount of thymine, and that the amount of guanine will always equal the amount of cytosine. However, the ratio of (adenine + thymine) to (guanine + cytosine) may vary from species to species. This feature is exploited in bacterial taxonomy (classification). DNA can be extracted from bacterial cells, purified, and the relative amounts of the four bases determined. The content of (guanine + cytosine) expressed as a proportion (or percentage) of the total base content of the DNA is called the *GC ratio* of that DNA. Species of bacteria which differ widely in their GC ratios are considered to be distantly related taxonomically; however, close GC ratios do not *necessarily* indicate close relationships, although they may do so if the

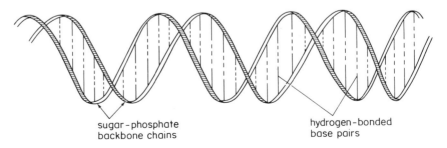

sugar-phosphate backbone chains

hydrogen-bonded base pairs

Fig. 6-6 The DNA double helix.

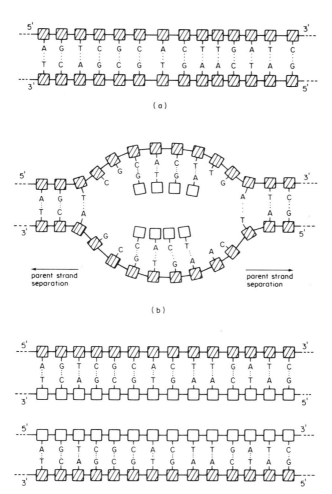

Fig. 6-7 DNA synthesis (replication): an outline of the principles. (The details of DNA synthesis are complex; the principles illustrated here are sufficient for an understanding of the role of DNA as discussed in this chapter.) (a) The "parent" DNA duplex. (b) The two strands of the parent duplex unwind and separate — separations proceeding in both directions as indicated. The exposed bases of the parent strands pair with complementary bases of free nucleotides, and bonds are formed between the latter by an enzyme called DNA polymerase. (c) The two new DNA duplexes each contain one new strand and one of the original parental strands — ie. replication is *semi-conservative*. (A = adenine, C = cytosine, G = guanine, T = thymine.)

bacteria show similarities in other respects. An example of the use of GC ratios can be found under *Planococcus* in Chapter 12.

We have not yet mentioned the order in which the bases occur in a strand of DNA. As it is the sequence of bases in DNA which holds the secret of the genetic code we shall be examining this in more detail later.

In contrast to DNA, *RNA* is always single-stranded in bacteria. However, a strand of RNA may be extensively folded into a three-dimensional structure which is stabilized by hydrogen bonding between short sequences of bases in different parts of the strand. In RNA — which does not contain thymine — adenine pairs with uracil.

The synthesis of nucleic acids

DNA synthesis As mentioned above, the sequence of bases in DNA encodes the genetic information necessary for the organization and function of the cell. When a cell divides each daughter cell must receive an accurate copy of this information. Therefore, prior to cell division, the sequence of nucleotides in the chromosomal DNA must be duplicated exactly in the synthesis of a second chromosome. This is achieved by a mechanism which exploits the specificity of base-pairing between nucleotide bases. During the process of DNA synthesis each strand of the original DNA duplex acts as a pattern or *template* on which a new strand is synthesized; this process is outlined in simplified

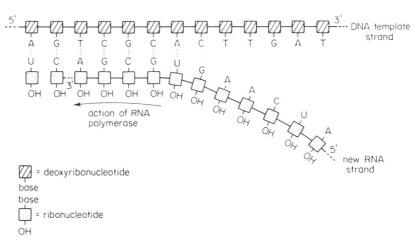

Fig. 6-8 Transcription: the synthesis of an RNA strand on a DNA template strand. (A = adenine, C = cytosine, G = guanine, U = uracil, T = thymine.)

form in Fig. 6-7. DNA synthesis can begin only at a particular base sequence (*initiation site*) in the DNA; any DNA molecule which lacks an initiation site cannot be replicated by the cell.

RNA synthesis RNA synthesis is similar, in principle, to DNA synthesis — although in this case the new RNA strand is synthesized on a *DNA* template and not on a parental RNA strand. Thus, base pairs are formed between the bases of deoxyribonucleotides in a DNA strand and the bases of free ribonucleotides (Fig. 6-8). An enzyme called RNA polymerase connects the ribonucleotides to give an RNA strand with a base sequence complementary to that of the DNA template. (Since RNA does not contain thymine, uracil pairs with adenine on the DNA template.) The newly-formed RNA strand peels away from the DNA template, allowing the DNA duplex to re-form. The synthesis of RNA on a DNA template is called *transcription.*

Protein synthesis and the genetic code

All bacterial activities are catalysed and regulated by specific proteins (enzymes), and proteins also play other regulatory and structural roles in the cell. Hence the genetic 'instructions' encoded in the DNA of a cell can dictate the nature of that cell by governing the synthesis of proteins. In other words, protein synthesis is the process in which the genetic code is 'decoded'.

A protein consists of one or more *polypeptides* — a polypeptide being a chain of amino acids linked together by peptide bonds ($-CO.NH-$). A polypeptide is folded or coiled into a three-dimensional structure which is stabilized by bonds (such as hydrogen bonds or disulphide bonds) formed between amino acids in different parts of the chain. This three-dimensional structure is essential for the biological activity of the polypeptide and is determined by the nature and number of its constituent amino acids and by the sequence in which they occur. The nature, number and sequence of amino acids in a polypeptide is dictated by the sequence in which the bases occur in a particular part of a DNA strand; a sequence of bases which dictates the structure (and hence function) of a polypeptide is called a *gene.*

Proteins, unlike nucleic acids, cannot be assembled directly on a DNA template: amino acids cannot interact with the bases on a strand of DNA. Hence protein synthesis must involve a number of intermediate stages. In the first stage the base sequence of a gene is *transcribed* (Fig. 6-8) to form a strand of *messenger RNA* (mRNA),

70

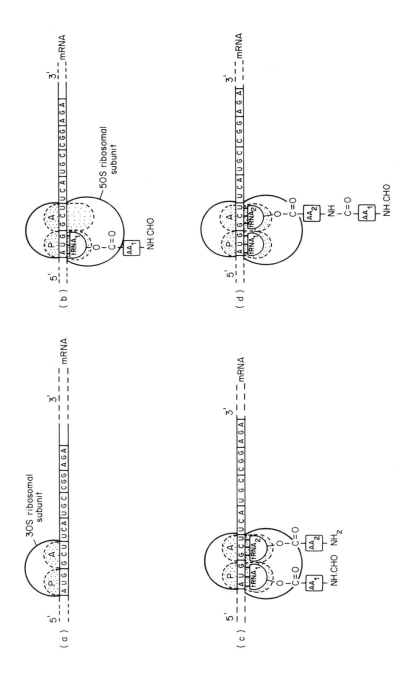

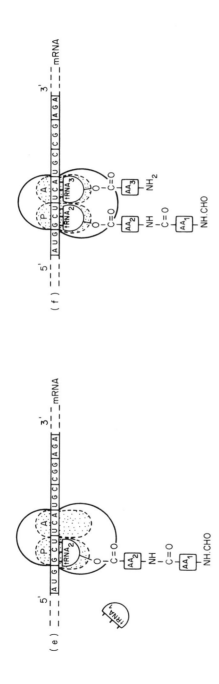

Fig. 6-9 Protein synthesis in bacteria. (a) the 30S ribosomal subunit binds to mRNA at a binding site near the beginning of a gene. (b) The first amino acid, AA_1 (= N-formylmethionine), linked to its tRNA ($tRNA_1$), occupies a site designated the *P site* on the ribosome; the anticodon of $tRNA_1$ is thus positioned opposite the initiator codon AUG. The 50S ribosomal subunit binds to complete the 70S ribosome. (c) The second amino acyl-tRNA binds to a site on the ribosome called the *A site*, so that its anticodon lies opposite the second codon. (d) The first amino acid is transferred to the second and a peptide bond is formed between them. (e) The ribosome moves along the mRNA by one codon, and $tRNA_1$ is ejected. The dipeptidyl-$tRNA_2$ thus comes to occupy the P site. The vacant A site now lies opposite the third codon. (f) The third amino acyl-tRNA binds at the A site. Steps (d) to (f) are repeated for each codon on the mRNA until a termination codon is reached; the ribosome then releases the mRNA and the completed polypeptide chain.

When the first ribosome has travelled along the mRNA, a second ribosome can bind and begin translation, followed by a third, and so on. Thus any mRNA molecule usually has many ribosomes at various positions along its length, each synthesizing a molecule of polypeptide.

and it is this mRNA which functions as the 'guide' for the assembly of amino acids. However, amino acids cannot interact with mRNA any more than they could with DNA. Each amino acid must first be linked to a specific adaptor or carrier molecule which can recognize both that amino acid and specific regions on the mRNA; these carrier molecules consist of another type of RNA called *transfer RNA* (tRNA). An amino acid is linked to its corresponding tRNA molecule (to form an amino acyl-tRNA) in a reaction which requires an input of energy in the form of ATP.

We are still left with a problem: how does the mRNA molecule 'line up' the amino acyl-tRNA molecules in the correct sequence? In other words we have still to crack the genetic code. We know that the code transcribed from DNA to mRNA has only four 'letters' or bases: A (adenine), U (uracil), G (guanine), and C (cytosine). Since twenty different amino acids occur in proteins, the code clearly cannot work on the principle that a single base specifies a single amino acid. An alternative possibility is that a short sequence of bases codes for each amino acid, ie. two or more 'letters' (bases) on the mRNA molecule make up 'words' (*codons*); thus each codon could specify a particular amino acyl-tRNA, and the sequence of codons in mRNA could specify the sequence of amino acids in a polypeptide. So, how many bases are there in a codon? Two bases per codon would allow the formation of only 16 (4^2) different codons, so that each codon must have more than two bases. If a codon has three bases the number of possible codons (using the four different bases) is 64 (4^3). This is, in fact, the case: each amino acid in a polypeptide is specified by a codon which consists of a particular sequence of three bases in a strand of mRNA. Since there are 20 amino acids to code for, and 64 codons are theoretically possible, either some codons must have no function or at least some amino acids must be coded for by more than one codon. In fact the latter is true. The function of each of the possible 64 codons is now known, and it has been found that some amino acids have as many as six different codons; for example, leucine is coded for by any of the following codons: UUA, UUG, CUU, CUC, CUA, and CUG. (As we shall see later, three of the 64 possible codons do not code for amino acids at all.)

Now, how does an amino acyl-tRNA recognize the correct codon on the mRNA? Each tRNA molecule contains a particular group of three bases, called an *anticodon,* which is complementary to the codon that specifies the amino acid it carries; thus, an amino acyl-tRNA becomes attached to the mRNA by base-pairing between the appropriate codon on the mRNA and the anticodon on the tRNA.

The actual linkage of amino acids to form a polypeptide requires the participation of ribosomes (Chapter 2). The details of the process are complex. Essentially, a ribosome binds to mRNA at the beginning of a gene and moves along the mRNA by one codon at a time, aligning amino acyl-tRNAs correctly with their codons and forging peptide bonds between the amino acid residues as it goes (Fig. 6-9). When it reaches the end of the gene the ribosome detaches from the mRNA and the completed polypeptide is released. The synthesis of a polypeptide upon mRNA in this way is called *translation.*

As we shall see in the following section, a single molecule of mRNA may carry sufficient information for making two or more different polypeptides: ie. it may carry transcripts of two or more genes. Clearly translation must start and finish at the beginning and end of each gene on the mRNA. Additionally, the codons themselves can make sense only if translation begins at the correct starting point — ie. the message must be read *in phase.* The genetic message must therefore be punctuated with 'start' and 'stop' signals. The 'start' signal, or ribosome binding site, positions the ribosome on the mRNA so that an *initiator codon,* AUG, lines up with the ribosomal P site — see Fig. 6-9. AUG normally codes for methionine. However, when it functions as an initiator codon, AUG binds to a special tRNA carrying N-formylmethionine, so that this amino acid is the first to be incorporated in any polypeptide. (However, no bacterial polypeptide has N-formylmethionine at one end; the formyl group — or one or more amino acid residues — may be removed by enzymes before the polypeptide becomes functional.) A 'stop' signal may be one of three different codons, called *termination codons* (or *nonsense codons*), and these are UAA, UAG, and UGA. These are the three codons, mentioned earlier, which do not code for amino acids. When a ribosome reaches a termination codon it becomes detached from the mRNA, and polypeptide synthesis is thus terminated.

A remarkable feature of the genetic code is that it is apparently universal: all organisms, whether prokaryotic or eukaryotic, appear to use exactly the same code. We can only guess at how such a code arose during the evolution of living organisms.

The control of gene expression

In the previous section we discussed the way in which a gene is transcribed into mRNA which, in turn, is translated into a polypeptide. However, at any given time not all of the genes in a cell are being transcribed and translated. Clearly, it would be wasteful of energy and materials to synthesize eg. an enzyme which is not needed under the prevailing environmental conditions. In fact a cell can control the synthesis of a particular enzyme by controlling the transcription of the appropriate gene. In this way the production of an enzyme may be 'turned on' (*induced*) in the presence of its substrate, or 'turned off' (*repressed*) in the presence of one of its products. Usually, inducible enzymes are responsible for the breakdown of substrates, while repressible enzymes are concerned with biosynthetic reactions.

In many cases a cell can co-ordinate the transcription of several genes which code for functionally related products — such as the enzymes involved in a particular metabolic pathway; a group of genes whose transcription is under common control is called an *operon*. The mechanisms of control are varied and often complex. We shall consider as an example the *lac* operon of *Escherichia coli;* this operon contains genes which code for enzymes concerned with the degradation of lactose, and is an example of an inducible system. The *lac* operon is probably understood more thoroughly than any other, but it should not be taken to be representative of operons in general: at least some other operons involve different control mechanisms.

The *lac* operon in *Escherichia coli* Lactose is a disaccharide consisting of glucose and galactose. Before it can be used as a nutrient it must be taken up by the cell and split into its component monosaccharides by an enzyme called ß-galactosidase. In the *lac* operon one of the genes (the Z gene) specifies the synthesis of ß-galactosidase, while another (the Y gene) specifies the synthesis of a protein ('galactoside permease') which is concerned with the transport of lactose across the cell membrane. A third gene, the A gene, specifies an enzyme called thiogalactoside transacetylase — an enzyme whose function is unknown. The Z, Y and A genes are called *structural genes.*

On the chromosome of *E. coli* the genes Z, Y and A occur in sequence; nearby is a *regulator gene,* designated I (see Fig. 6-10). P is a site called the *promoter* to which the enzyme RNA polymerase binds prior to transcription of the structural genes. The regulator gene, I, is transcribed and translated independently and continuously, and the product is a protein called the *repressor protein.* When no lactose is present, this repressor protein binds to a region of DNA called the *operator, O,* and effectively blocks the path of the RNA polymerase

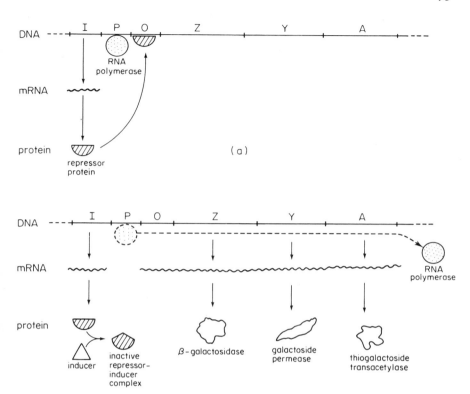

Fig. 6-10 The *lac* operon of *Escherichia coli*. (a) In the absence of an inducer the repressor protein product of the *I* gene binds to the operator, *O*, and prevents transcription of the structural genes *Z, Y* and *A* by RNA polymerase. (b) When an inducer is present it binds to the repressor, preventing it from attaching to the operator. Transcription of *Z, Y* and *A* proceeds, followed by translation of the three enzymes.

bound to the promoter, *P* (see Fig. 6-10); transcription of the *Z, Y* and *A* genes is therefore prevented.

How then does the presence of lactose reverse this blockade imposed by the repressor? Lactose itself does not affect the repressor. However, an enzyme within the cell converts some of the lactose' to *allolactose*, and this is the molecule responsible for lifting the blockade — ie. allolactose is the *inducer* of the *lac* operon. The inducer binds to the repressor protein and changes it in such a way that it releases its hold on the operator. With the repressor protein removed the RNA polymerase is free to begin transcription of the *Z, Y* and *A* genes. Transcription proceeds to the end of the operon, resulting in the formation of a single strand of mRNA which contains a transcript of all three structural genes; as the promoter becomes vacant, a second polymerase molecule can bind and begin transcription, and so on. Each of the three genes on each mRNA strand is translated, as discussed in the previous section, and the cell can then begin to metabolize lactose.

When all the available lactose has been metabolized, transcription of the *lac* operon is no longer required and the repressor blockade must be re-established. As the level of lactose falls, fewer molecules of inducer are available to bind to the repressor protein; molecules of free repressor therefore become available to attach once more to the operator, so that transcription of the *lac* genes is turned off. However, what about the mRNA molecules already synthesized? Could not translation continue indefinitely on these? In fact this does not happen because in bacterial cells mRNA is short-lived, and protein synthesis can continue only for so long as new mRNA molecules are being produced.

Catabolite repression If *Escherichia coli* is supplied with both glucose and lactose, it might reasonably be expected that the synthesis of *lac* enzymes would be induced so that lactose as well as glucose could be metabolized. In fact this does not happen. Despite the presence of lactose — and hence the allolactose inducer of the *lac* operon — the synthesis of *lac* enzymes remains repressed until all the glucose has been metabolized. This is an example of a phenomenon known as *catabolite repression* or the *glucose effect* — of which diauxie (Chapter 3) is a manifestation. Catabolite repression occurs quite commonly in bacteria and means that if a range of substrates is available some can be used in preference to others. The mechanism of catabolite repression is complex and may differ from one system to another. It is mentioned here merely to give an indication of the intricate nature of the control mechanisms which integrate the various aspects of metabolism in a bacterial cell.

MECHANISMS FOR GENETIC CHANGE

We have seen that the synthesis of a new strand of DNA involves the exact replication of an existing strand, and that when a cell divides each daughter receives an accurate copy of the original chromosome. Does this mean that the genetic information in a bacterium can never change? If this were so all cells of a given bacterial species would be identical and would never show any variation — no matter how many cell divisions were to occur. In fact the genetic system is much more flexible than this. The genetic information in a bacterium can be altered or supplemented in two basic ways: (i) by changes in chromosomal DNA as a result of mutation, and (ii) by the introduction into a cell of new genetic information in the form of a piece of DNA from another source.

Mutation

In a bacterium, a mutation is a stable and heritable change in the sequence of nucleotides in its DNA. Since the sequence of nucleotides in DNA encodes genetic information, any change in that sequence will usually alter the information carried by the DNA; hence, a mutation in a particular gene will normally alter the product of that gene. Mutations occur spontaneously at a very low rate; for example, in a population of bacterial cells a mutation affecting a particular gene may occur at a frequency of about once in every 10^6-10^{10} cell divisions. The rate at which mutations occur can be increased in the laboratory by treating the cells with certain physical or chemical agents known as *mutagens*. Physical mutagens include some types of radiation, eg. X-rays and ultraviolet radiation. Chemical mutagens include eg. substances such as alkylating agents and nitrous acid (which react directly with DNA) and chemical analogues of the nitrogen bases in DNA. Mutagens function in a variety of ways. For example, chemical alteration of certain nitrogen bases in DNA may alter their base-pairing properties, and hence errors will be introduced during the subsequent round of DNA replication; such errors may include eg. the substitution of one base for another.

Mutations occur at random, affecting different genes in different individuals throughout a population. A mutation is often harmful to the cell in which it occurs, and may even be lethal — as, for example, when the gene in which the mutation occurs normally codes for a product essential to the cell. Nevertheless, some mutations can be beneficial; for example, a mutation may give a cell increased resistance to an antibiotic or other toxic agent. A mutation which is advantageous to a cell may well confer on that (*mutant*) cell an increased fitness for the prevailing environmental conditions, and such a mutant may outgrow the other (*wild-type*) members of the population — eventually becoming the dominant cell type. Such 'natural selection' is, of course, the basis of the modern concept of evolution.

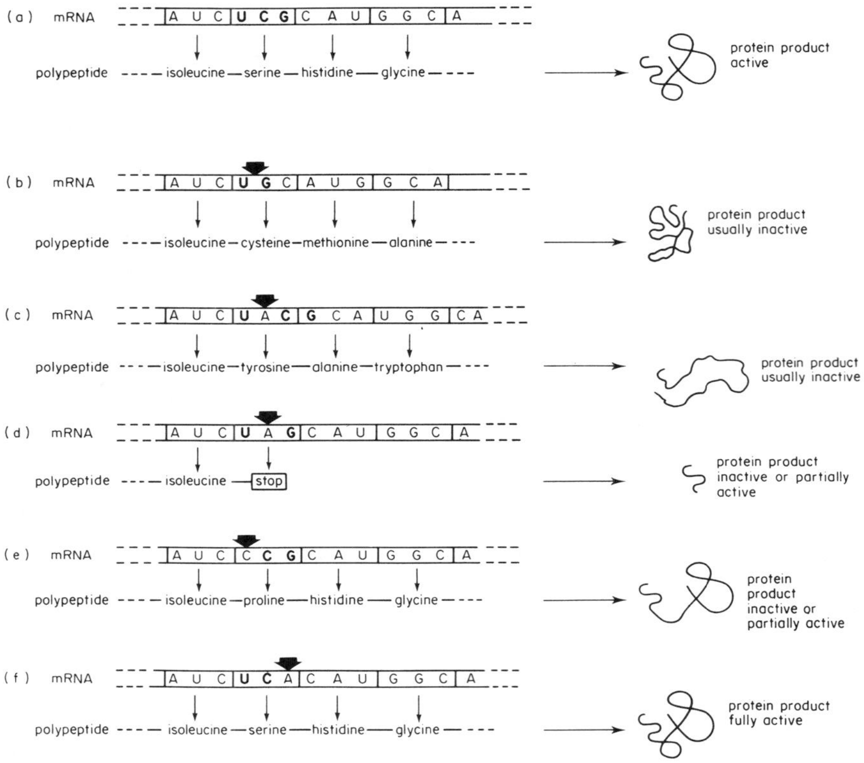

Fig. 6-11 Point mutations and their consequences. For ease of comparison all mutations are shown to occur in the serine codon UCG. The site of the point mutation is indicated by a heavy arrow, ➧. (a) mRNA and polypeptide synthesized from the normal (wild-type or non-mutant) gene. (b) *Deletion* from codon UCG of a single nucleotide, C. The genetic message becomes out of phase and all codons subsequent to the deletion are altered. Such a mutation is called a phase-shift or frame-shift mutation*. (c) *Addition* to codon UCG of a single nucleotide, A. As in (b), this results in a phase-shift mutation*. (d) *Nonsense mutation*: a codon specifying an amino acid (ie. UCG) is altered to a chain-terminating codon (UAG) by base substitution (C → A). Protein synthesis therefore stops prematurely*. (e) *Mis-sense mutation*: a codon specifying one amino acid (UCG for serine) is altered by base substitution (U → C) so that it specifies a different amino acid (CCG for proline). The activity of the gene product depends on the nature of the incorrect amino acid and its position in the protein. (f) *Silent mutation*: a codon specifying one amino acid (UCG for serine) is altered by base substitution (G → A) to another (UCA) which codes for the same amino acid (serine). Such a mutation will have no effect on the activity of the gene product.

*The effect of such a mutation depends on its position in the gene; if it occurs near the end of the gene — so that most of the polypeptide is synthesized normally — the product may have some activity.

Types of mutation Mutations may involve any of several types of change in the nucleotide sequence in DNA. Occasionally whole regions of the DNA may be affected — sections may be lost, gained, reversed, etc; however, in many mutations a single nucleotide is lost or gained, or substituted by an alternative nucleotide. Some of the consequences of such *point mutations* are shown in Fig. 6-11.

The isolation of mutant bacteria A particular type of mutant cell will arise spontaneously in a population of bacteria only at very low frequency, and even if the population is treated with a mutagen any particular mutant will still be far outnumbered by wild-type cells and other types of mutant. Thus if we require a pure culture of a particular mutant bacterium we are faced with the task of singling out one (or a few) cells from among many millions. This is often not as difficult as it might seem. Suppose, for example, we wish to isolate a mutant which has acquired the ability to resist the effects of an antibiotic (eg. penicillin). A solid medium containing penicillin is inoculated with a large number of bacterial cells and is then incubated in the usual way. If the inoculum contains any mutants which are resistant to penicillin these cells will grow and each will form a colony; all the wild-type cells, which are sensitive to penicillin, will be inhibited by the antibiotic. This type of selection can be used whenever the mutant sought can grow under conditions which inhibit the growth of wild-type cells.

In some cases the isolation of a mutant is not so straightforward. This is true, for example, of a mutant which has lost the ability to synthesize an essential compound — eg. a particular amino acid — and which must therefore be supplied with that compound in order to grow; such a mutant is called an *auxotroph,* and the corresponding wild-type cell is termed a *prototroph.* In this case the problem is to separate the occasional auxotrophic mutant from a population of prototrophic cells. Clearly, selective media cannot be used since the prototrophs will be able to grow on any medium which allows the auxotrophs to grow.

The isolation of an auxotroph commonly exploits the inability of such a mutant to grow on a *minimal medium,* ie. a medium containing the minimum number of nutrients required for the growth of a prototroph. If such a (solid) medium is supplemented with a *low* concentration of a possible growth requirement — eg. the amino acid histidine — it will allow limited growth of histidine-requiring auxotrophic mutants as well as the normal growth of prototrophs. The colonies of auxotrophs soon exhaust the supply of histidine in their

80

vicinity and cease to grow, so that they remain small; colonies of proto-trophic cells attain their normal size. Small colonies are therefore presumed to be those of auxotrophs and can be subjected to further examination. In an alternative method for isolating auxotrophs, a low-density population containing both prototrophic and auxotrophic cells is inoculated onto a *complete medium,* ie. a medium which allows the growth of both prototrophs and auxotrophs. Following incubation, normal-sized colonies are formed by both types of cell, and in order to find out which colonies are those of the auxotrophs it is necesary to sub-culture each and every colony to a minimal medium; this mass sub-culture can be conveniently achieved by a process known as *replica plating.* In this method a disc of sterile velvet is attached to one end of a wooden cylinder of diameter equal to that of the plates used. The velvet is then pressed gently onto the surface of the complete medium (the master plate) on which are growing the colonies of both proto-trophs and auxotrophs. Cells from each of the colonies stick to the velvet — which is then pressed lightly onto the surface of a sterile plate of minimal medium (the replica plate). After the replica plate has been incubated a comparison is made between the positions of the colonies on the master plate and those on the replica plate; any colonies which appear on the master plate but not on the replica plate are presumed to be those of auxotrophs (Fig. 6-12).

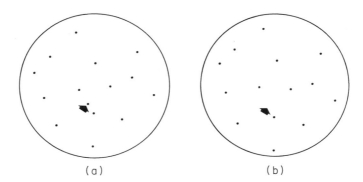

(a)   (b)

Fig. 6-12 Replica plating used as a method of the isolation of an auxotrophic mutant. (a) Master plate: complete medium with colonies of both prototrophs and auxotrophs. (b) Replica plate: minimal medium with colonies of prototrophs only. On the master plate an arrow indicates a colony of a presumed auxotrophic mutant; there is no colony at the corresponding position on the replica plate (arrowed) since the mutant cannot grow on minimal medium.

Gene transfer

A bacterium may obtain new genetic information by *transformation, transduction,* or *conjugation;* these processes differ from each other in the mechanism by which 'foreign' DNA enters the recipient cell. Since transduction requires the participation of bacteriophages we shall postpone a discussion of this process until Chapter 7 and consider here only transformation and conjugation.

Transformation Transformation is a process in which a bacterial cell takes up from its environment a fragment of DNA. The discovery of transformation followed the observation that when heat-killed cells of a virulent (disease-causing) strain of *Streptococcus pneumoniae* were mixed with live cells of a non-virulent strain, the latter became virulent; it was later found that the living cells had acquired virulence by taking up DNA released from the dead cells — indicating that this DNA carried information specifying the characteristic of virulence. This discovery was of tremendous importance in that it led to the recognition of the role of DNA as the carrier of genetic information in cells. Transformation is now known to occur in many other bacteria — eg. in species of *Bacillus, Haemophilus* and *Neisseria.*

In a given population of bacterial cells the ability to take up transforming DNA is not a permanent property of the cells, but rather depends on factors such as temperature, pH, the availability of certain nutrients, and the density of cells in the population. When a cell is able to take up DNA it is said to be *competent.* However, even a competent cell cannot take up *any* type of DNA fragment: transforming DNA must be double-stranded and of a molecular weight greater than a certain minimum value.

In the first stage of transformation DNA binds to the surface of a competent cell. (Cells which are not competent cannot bind DNA.) Binding is followed by the entry of DNA into the cell by a mechanism which is still largely unknown. In *Bacillus subtilis* and *Streptococcus pneumoniae* the DNA appears in the recipient cell in the form of a single strand; the complementary strand is apparently degraded — possibly during entry. Genes carried by the transforming DNA can be expressed in the recipient cell only if they are incorporated into the chromosome of that cell. Such incorporation is achieved by *recombination:* a process which appears to involve the replacement of part of one strand of the recipient's chromosome with part or all of the strand of transforming DNA.

Conjugation Conjugation in bacteria involves the transfer of DNA from one cell (the donor or 'male') to another (the recipient or 'female') while the two cells are in physical contact. Conjugation has been studied chiefly in Gram-negative bacteria; however, it has been discovered recently that a process which is apparently similar also occurs in species of the Gram-positive genera *Streptococcus* and *Streptomyces*.

Most of our knowledge of bacterial conjugation comes from studies on the process in *Escherichia coli*. The ability of an *E. coli* cell to act as a donor is due to the presence, in that cell, of a small piece of double-stranded DNA which is additional to that in the chromosome; this piece of DNA is known as a *plasmid*. A plasmid may occur in the cell as a small circular DNA molecule which is separate from the chromosome, or — in some cases — it may become integrated with the chromosome (see later). A plasmid contains an initiation site for DNA synthesis, which means that — when separate from the chromosome — it can be replicated within the cell; when integrated with the chromosome it is replicated as part of the chromosome. When the cell divides the plasmid is inherited by each daughter cell.

One well-studied plasmid is the *F plasmid* or *F factor* (F for fertility). The F plasmid carries genes which code for specific donor cell characteristics — such as the formation of appendages called sex pili (see Chapter 2). A donor cell which contains an F plasmid in the form of a separate circular molecule is designated F^+, while a recipient cell (which does not contain an F plasmid) is designated F^-. When F^+ and F^- cells are mixed contact is quickly established between donor and recipient cells; initially contact appears to be made through the sex pilus, although direct cell-to-cell contact may often — or perhaps always — follow (Fig. 6-13). Once effective contact has been established one (linear) strand of the double-stranded F plasmid is transferred from the donor to the recipient (Fig. 6-13). DNA replication must then occur in both cells in order to re-establish the double-stranded nature of the F plasmid, and the recipient must also connect the two ends of the linear plasmid DNA to restore its circular nature. In this way the F^- cell becomes an F^+ donor which can, in turn, conjugate with another F^- recipient cell.

An F plasmid occasionally loses its independence and becomes integrated with the chromosome (Fig. 6-14); in any population of F^+ cells this occurs in about 1 cell in every 10^5. This creates another type of donor cell which is designated *Hfr* (for high frequency of recombination — see later). Like F^+ donors, Hfr donors can conjugate with

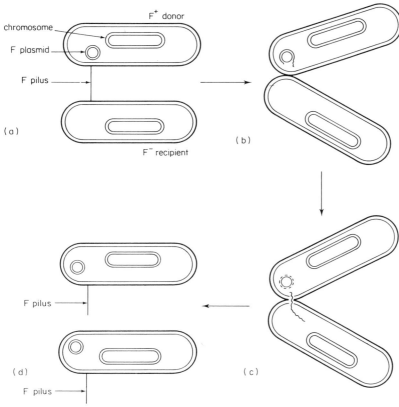

Fig. 6-13 Bacterial conjugation between an F^+ donor and an F^- recipient*. (a) Initial contact between an F^+ and an F^- cell is made via the sex pilus (F pilus). (b) Cell-to-cell contact is established. A break is made, by an enzyme, in one (or both) strands of the F plasmid. (c) A single strand of the F plasmid is transferred to the F^- cell. This strand is replaced by synthesis of a new strand in the donor, using the remaining strand as a template. (d) The cells separate, and a complementary strand is synthesized in the recipient cell. The cell which was formerly F^- is now F^+.

*The diagram shows a *tentative* scheme since many features of bacterial conjugation are as yet unknown.

F^- cells. During Hfr $\times$ F^- conjugation the transfer of DNA to the recipient is preceeded by a break in the plasmid DNA; consequently, part of the F plasmid is transferred first, but this is followed by the attached strand of chromosomal DNA (Fig. 6-14). Hence only after one strand of the entire chromosome has been transferred can the remainder of the F plasmid enter the recipient cell. Since the DNA strand usually breaks before its complete transfer has been achieved, the recipient in an Hfr $\times$ F^- cross usually does not receive a complete

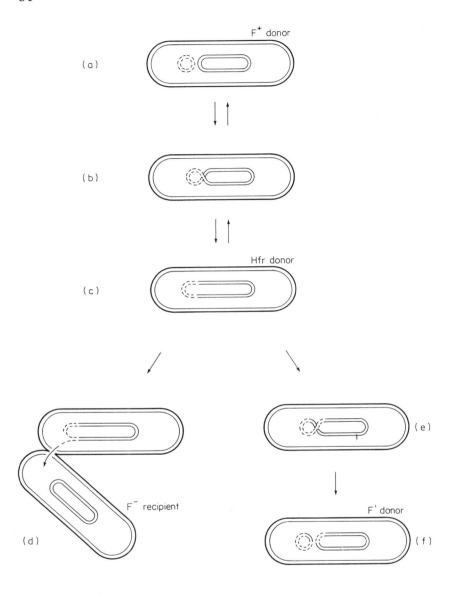

Fig. 6-14 The formation of Hfr and F′ donors. (a) An F⁺ donor cell. (b) The F plasmid integrates with the chromosome by the formation of a single cross-over between the two circular DNA molecules. (c) An Hfr donor results. (d) Conjugation between an Hfr donor and F⁻ recipient. *Part* of the F plasmid is transferred, followed by chromosomal DNA. (e)-(f) Aberrant excision of the F plasmid produces an F′ plasmid which includes some chromosomal DNA but which can still function as an F plasmid. An F′ donor can conjugate with an F⁻ recipient as shown in Fig. 6-13.

F plasmid and so remains F$^-$; however, on the rare occasions when the entire chromosome *and* F plasmid are transferred the recipient becomes an Hfr donor. In either case the recipient cell will usually receive chromosomal DNA from the donor cell, and if the donor genes can undergo recombination with the recipient's chromosome they will become part of the recipient's genetic constitution. The high proportion of such recombinant cells which results from Hfr $\times$ F$^-$ crosses, as compared with F$^+$ $\times$ F$^-$ crosses, is the reason why this type of donor is designated 'Hfr'.

The integration of the F plasmid with the bacterial chromosome is reversible — ie. the F plasmid may leave the chromosome and once more behave independently. In other words, F$^+$ cells may be produced from Hfr cells. Occasionally, however, as the F plasmid leaves its integrated position in the chromosome it takes with it a neighbouring portion of chromosomal DNA (Fig. 6-14(e) and (f)), and the resulting modified F plasmid is designated F$'$ (F-prime). Conjugation between an F$'$ donor and an F$^-$ recipient results in the transfer of the F$'$ plasmid — containing its portion of chromosomal DNA — to the recipient. Thus, F$'$ donors have characteristics intermediate between those of Hfr and F$^+$ donors: the ability to function as a donor is transferred (as in F$^+$ $\times$ F$^-$ crosses), and donor chromosomal DNA is transferred (as in Hfr $\times$ F$^-$ crosses).

The F plasmid is but one of many types of plasmid found in bacteria. Other types of plasmid code for functions such as eg. the synthesis of toxins (see Chapter 9) or the synthesis of enzymes which increase the metabolic potential of the cells which contain them. One important group of plasmids, the *R plasmids* (or *R factors*), confer on the cells which contain them an increased resistance to particular antibiotics and/or to certain other harmful agents (see Chapter 10).

Not all plasmids can bring about their own cell-to-cell transfer by conjugation; those which can are called *transmissible* or *conjugative* plasmids, while those which cannot are called *non-transmissible* or *non-conjugative* plasmids. A non-conjugative plasmid may be transferred from donor to recipient during conjugation promoted by a conjugative plasmid occupying the same donor cell; such a plasmid can also be transferred by transduction (see Chapter 7).

'Genetic engineering' with bacteria

Bacteria produce specific enzymes called *endonucleases* which break double-stranded DNA at sites which contain certain sequences of

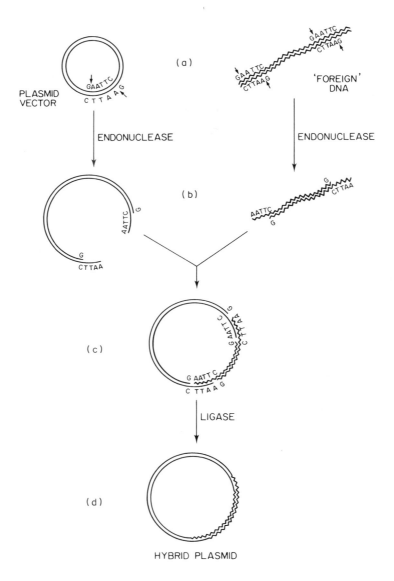

Fig. 6-15 "Genetic engineering" with bacteria: the construction of a hybrid plasmid. (a) A plasmid (isolated from a bacterium) and "foreign" DNA (eg. from a eukaryotic cell) are each treated with the same endonuclease which cleaves double-stranded DNA eg. in the positions shown by arrows. Endonuclease action produces fragments of double-stranded DNA with free single-stranded ends which are complementary (b). (c) Base-pairing can occur between the single-stranded ends of the two types of DNA fragment. (d) The broken ends are linked by means of a ligase to form a closed, double-stranded loop of DNA: the hybrid plasmid. (NB. The cloning vehicle may be bacteriophage DNA (see Chapter 7) instead of a plasmid.)

bases; other enzymes, called *ligases,* join together free ends of DNA strands. These two types of enzyme can be isolated and used to construct an artificial hybrid plasmid by incorporating in a natural plasmid (the *vector* or *vehicle*) a fragment of DNA carrying genes from any of a variety of prokaryotic or eukaryotic sources (see Fig. 6-15). The hybrid plasmid can then be introduced into a bacterial cell eg. by transformation; the sequence of bases necessary for the initiation of DNA replication is provided by the plasmid vector, so that the hybrid plasmid can replicate within the bacterial cell and will be inherited by successive generations of cells. In this way we can use bacteria to generate many copies of one or more specific genes — a technique known as DNA *cloning.* Cloning is a very useful technique for genetic experimentation — eg. it allows the production of quantities of genes and gene products sufficient for physical and chemical analysis. Additionally the technique is being developed on a commercial basis for the production of medically or industrially important compounds which are difficult or expensive to produce in other ways; such substances include eg. certain human hormones, and interferons (anti-virus — and possibly anti-cancer — substances which are produced by animal cells under certain conditions). This is a new and rapidly developing branch of 'biotechnology'; in some successful experiments to date (1980) genes coding for insulin, for a human growth hormone, and for an interferon have each been cloned in strains of *Escherichia coli* and products with apparently normal activity have been obtained.

The introduction of foreign DNA into a bacterium may be accompanied by certain risks: the bacterium may acquire, unpredictably, new and undesirable properties — eg. a new or enhanced capacity to cause disease. Such experiments are therefore strictly controlled and various safety precautions are taken; for example, bacteria used for cloning may be 'metabolic cripples' which can be grown only under special laboratory conditions and which could not survive if released into the environment.

7 Bacteriophages

Most or all bacteria can be infected by specialized viruses called *bacteriophages* (often abbreviated to *phages*). A virus is not an organism in the usual sense of the word: it does not have a cellular structure and cannot, by itself, metabolize or reproduce. However, once inside a living cell a virus may bring about its own replication by re-directing the synthesizing machinery of the cell to make virus components; the newly-formed viruses then leave the cell and can infect other cells.

Many different types of phage are known. In most cases a particular phage can infect the cells of only one genus, species, or strain of bacterium. An individual phage may be filamentous, polyhedral, or pleomorphic in shape, and in many cases there is a 'tail' by means of which the phage attaches to a bacterial cell. (An example of a polyhedral tailed phage is shown in Fig. 7-1.) Many phages consist only of nucleic acid and protein, the nucleic acid being enclosed by a protein coat called a *capsid*. According to phage, the nucleic acid may be double-stranded DNA, single-stranded DNA, or single-stranded RNA; there is also at least one phage (which attacks *Pseudomonas* species) which contains double-stranded RNA.

A phage can affect its bacterial host in one of two principal ways: either it invariably lyses (disrupts and kills) the host cell (in which case it is said to be *virulent*), or it can establish a stable relationship with the host cell (when it is said to be *temperate*).

Virulent phages: the lytic cycle

Fig. 7-2 illustrates in outline the lytic cycle of bacteriophage T4, a virulent phage whose host is *Escherichia coli*. The lytic cycles of other virulent phages may differ in detail. For example, among tailed phages

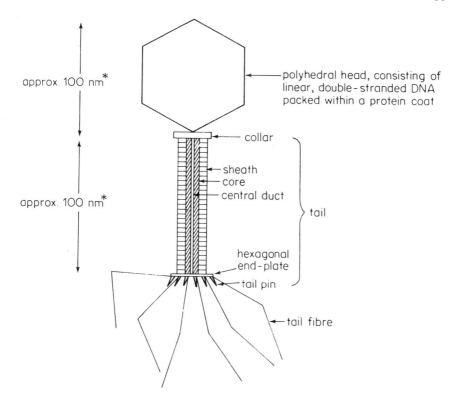

approx. 100 nm*

polyhedral head, consisting of linear, double-stranded DNA packed within a protein coat

collar

sheath
core
central duct

approx. 100 nm*

tail

hexagonal end-plate

tail pin

tail fibre

Fig. 7-1 Bacteriophage T4: a structurally complex bacteriophage which attacks *Escherichia coli.* The jointed tail fibres are shown extended; however, it appears that they are normally wrapped around the tail until the phage makes contact with a host cell (Fig. 7-2).

*1nm (nanometre) = 1 millionth of a millimetre $(10^{-6}$ mm).

the tail is not always contractile; additionally, some virulent phages do not bring about the immediate destruction of the host's chromosome since they can replicate only with the help of certain of the host's own genes. Not all phages attach to the bacterial cell wall; some attach specifically to flagella, while others attach to certain types of pilus. In RNA-containing phages the RNA functions as mRNA in the host cell. These phages must bring about the replication of their own RNA since no bacterial enzyme can synthesize RNA on an *RNA* template. Some small RNA phages bring about the formation of an RNA replicase which contains four protein subunits — one of which is coded for by a phage gene while the other three are 'borrowed' from the protein-synthesizing machinery of the host cell.

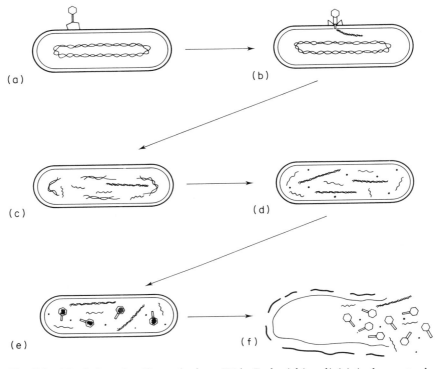

Fig. 7-2 The lytic cycle of bacteriophage T4 in *Escherichia coli*. (a) A phage attaches by its tail fibres to specific "receptor sites" on the cell wall of a host cell. Subsequently the tail makes contact with the cell wall. (b) The phage sheath contracts and the inner core penetrates the cell wall. DNA passes from the phage head to the cell through the central duct. The empty phage capsid remains at the cell surface. (c) Part of the phage DNA (the "early genes") is transcribed into mRNA, which is in turn translated using the host cell ribosomes. One of the "early" proteins is an enzyme which breaks down the host cell chromosome. (d) Phage DNA is replicated. Futher transcription of phage genes (the "late genes") results in mRNA molecules on which are synthesized structural phage proteins (protein components of the head, tail etc — shown in the diagram as black dots). (e) Phage DNA becomes condensed and associates with head proteins, and new phage particles are assembled. (f) A phage-coded enzyme, lysozyme, attacks the peptidoglycan of the host cell wall. The weakened wall allows osmotic lysis of the cell, so that the newly-matured phage particles (numbering 100 or more) are released.

If virulent phages are added to a broth culture of susceptible bacteria, phage replication will bring about the lysis of most or all of the cells; cell lysis on such a massive scale can cause a dense, cloudy culture to become clear. The effects of a virulent phage can also be seen on bacteria growing on a solid medium. If a small number of virulent phages is added to a layer of confluent bacterial growth, each phage infects a cell and causes it to lyse, releasing many phages which can then infect and lyse neighbouring cells; in this way a visible, usually circular clearing — called a *plaque* — is formed in the opaque layer of confluent growth.

Virulent phages are sometimes a considerable nuisance in certain industrial processes which involve bacteria — processes such as cheese-making and antibiotic production.

Temperate phages: lysogeny

A temperate phage is one which can enter into a stable relationship with a bacterium; this relationship is called *lysogeny*, and the bacterial host is said to be *lysogenic*. A lysogenic bacterium can live and grow more or less normally, and is immune from attack by other phages of the same type. Phage replication is co-ordinated with host cell division so that the phage is inherited by each daughter cell. However, under certain conditions the lysogenic relationship breaks down and the phage initiates a lytic cycle, destroying its host; thus, any lysogenic cell is potentially capable of releasing phage progeny. Lysogeny appears to be a common phenomenon; a high proportion of bacteria isolated from natural environments are found to be lysogenic.

The best-known temperate phage is phage *lambda* — a phage which infects *Escherichia coli*. Phage *lambda* contains linear, double-stranded DNA in a polyhedral head, and has a non-contractile tubular tail with a single terminal tail fibre. On entry into a host cell, the phage DNA becomes circular and transcription of phage genes begins. At this stage either a lytic cycle or lysogeny may follow — depending largely on environmental conditions. In the establishment of lysogeny a phage gene codes for a repressor protein which prevents transcription of those phage genes responsible for bringing about the lytic cycle. Another phage gene product enables the phage DNA to insert into a specific site in the bacterial chromosome — apparently in a manner similar to that in which the F plasmid inserts into the chromosome to form an Hfr donor (see Fig. 6-14). Once integrated, the phage DNA — now referred to as a *prophage* — is replicated as part of the chromosome every time the host cell divides.

Lysogeny can continue for as long as the phage repressor prevents transcription of those genes which control the lytic process. If, for any reason, repressor ceases to be produced — or is inactivated — the lytic process is initiated. (Certain mutants of phage *lambda* produce a defective repressor, and these mutant phages are always virulent.) In the *E. coli*/phage *lambda* system a change from the lysogenic to the lytic state can be brought about experimentally by subjecting a lysogenic population of cells to certain agents — including eg. ultraviolet radiation or certain chemicals; this is known as *induction,* and may involve the inactivation of the phage repressor.

The *E. coli*/phage *lambda* system may not be representative of lysogeny in general. For example, a temperate phage known as P1 (which also infects *E. coli*) appears not to integrate with the chromosome of its host cell during lysogeny.

Phage conversion

Bacteria which are infected by phage may show certain characteristics not exhibited by uninfected cells; this phenomenon may be due eg. to the expression of phage genes in the cells and is called *phage conversion.* For example, the strains of *Corynebacterium diphtheriae* which cause diphtheria contain the phage *beta* and produce a potent toxin (see Chapter 9); the gene coding for this toxin occurs in the phage DNA and not in the bacterial chromosome. Strains of *C. diphtheriae* which are not infected with the *beta* phage cannot produce toxin and cannot cause diphtheria.

Transduction

We are now in a position to examine the third method of genetic transfer mentioned in Chapter 6 — namely, *transduction.* Transduction involves the transfer of bacterial genes from one cell to another via a bacteriophage. There are two types of transduction: *generalized* transduction and *specialized* transduction.

Generalized transduction In generalized transduction any of a variety of genes in one bacterium may be transferred to another. In a population of bacteria infected with phage, it may occasionally happen that — during phage assembly — a phage head incorporates a short piece of double-stranded chromosomal DNA in place of the usual phage DNA. Thus, when the phage progeny are released a very small proportion will contain pieces of bacterial chromosome. Such abnormal phages can attach to other bacterial cells and inject their DNA in the usual way; however, since this DNA is not phage DNA neither lysogeny nor cell lysis will result.

The fragment of chromosomal DNA thus introduced into the recipient cell can undergo one of two possible fates. In most cases the fragment is *not* integrated into the recipient's chromosome. The genes carried by the fragment may be expressed (ie. transcribed and translated), but usually the fragment cannot be replicated. Thus, when the recipient cell divides, only one of the daughter cells can receive the DNA fragment, and in the population which results from subsequent cell divisions only one cell will contain the transduced DNA fragment. This is called *abortive transduction.* Alternatively, the transduced fragment may undergo recombination with the recipient's chromosome so that its genes become a permanent part of the recipient cell's genetic constitution; this is *complete transduction.*

Specialized transduction Specialized transduction (also called *restricted transduction*) can be brought about by a temperate phage such as *lambda.* When a *lambda*-lysogenized cell is induced, the prophage leaves the chromosome in much the same way as an F plasmid leaves an Hfr chromosome — see Chapter 6. Just as an F plasmid may, during excision, take with it some chromosomal genes to form an F′ plasmid (see Fig. 6-14), so a *lambda* prophage may, occasionally, take with it adjacent chromosomal genes. The *lambda* prophage normally fits into the chromosome of *Escherichia coli* at a site between genes designated *gal* (for galactose utilization) and *bio* (for biotin synthesis). Thus, on induction a small proportion of the progeny phages may contain either *gal* or *bio* genes — usually at the expense of some phage genes which are left behind in the bacterial chromosome. When such defective phages infect other (recipient) cells the genes *gal* or *bio* will enter the recipient cells along with the remaining phage genes.

8 Bacteria in the living world

Bacteria are often thought of either as pests to be destroyed, or as convenient 'bags of enzymes' useful for experimental purposes. However, bacteria have a life of their own outside the laboratory, and many of their activities have important repercussions in the natural environment. This aspect of bacteriology has many facets, and in this chapter we can give only a brief outline of some of them.

Relationships between bacteria and other organisms

Microbial communities Most bacteria are *free-living*, ie. they do not necessarily form specific associations with any other organisms; nevertheless, they are part of the web of life, and in nature they can rarely grow without affecting — or being affected by — other organisms. Normally bacteria occur as members of mixed communities which may include fungi, algae, protozoa and other organisms. Microbial communities such as these can be found in almost any natural habitat: in soil, in water, on the surfaces of plants, on and within the bodies of man and other animals, etc. The microorganisms which are normally present in a particular habitat are referred to, collectively, as the *microflora* of that habitat.

Any unoccupied microhabitat will rapidly become colonized by microorganisms which may affect one another in various ways. For example, they may have to compete for scarce nutrients, for oxygen (in aerobic habitats), for space etc, and those organisms which cannot compete effectively are soon eliminated from the habitat. In some cases an organism can actively discourage at least some of its competitors by producing substances toxic to them — a phenomenon which may be termed *antagonism;* in nature the ability of a microorganism to produce antibiotics (Chapter 10) probably gives it a competitive advantage. There may also be relationships between microbial colonists in which one or both organisms benefit and neither organism is harmed; for example, an acid-producing organism will help to create conditions favourable to a species whose growth requires a low pH.

If the habitat being colonized remains undisturbed there will eventually develop a stable community of organisms in which the various beneficial and detrimental interactions have reached a delicate state of balance. An alien microorganism will often have difficulty in establishing itself in such a community — unless a disturbance in the environment upsets the balance in the community. For example, in the intestine of an animal the natural microflora can often discourage the establishment of a pathogenic microorganism; however, if the micro- flora is disturbed — eg. as a result of antibiotic therapy — a pathogen may be able to become established and cause disease. Similarly, a plant pathogen which could not compete with a soil's microflora may be able to flourish in soil which has been subjected to anti-microbial treatment.

Nutritional relationships Organisms which obtain their nutrients from dead organic matter are called *saprotrophs* (or *saprophytes*). Some saprotrophic bacteria can use only soluble organic compounds which can be taken directly into their cells; usually such compounds can be used by a wide range of microorganisms, so that competition for them might be expected to be fierce. On the other hand, insoluble macromolecules (such as cellulose) cannot be taken up, directly, by any bacterial cell. Nevertheless, certain saprotrophic bacteria can secrete enzymes which degrade such compounds outside the cell, releasing soluble, assimilable products — a proportion of which will become available to other organisms in the vicinity. In some cases a complex organic substrate may be degraded in a stepwise manner by several saprotrophs acting sequentially, each species carrying out one or a few steps in the breakdown process. Such co-operation is very important in the degradation — and hence re-cycling — of organic compounds in nature, and there are very few biological compounds which cannot readily be broken down by a community of saprotrophic microorganisms.

In contrast to the saprotrophs, many bacteria secrete enzymes which lyse certain other (living) bacteria and fungi, thereby releasing soluble products which are used as nutrients; such bacteria include members of the order Myxobacterales: Gram-negative bacilli which occur in soil and plant litter. The myxobacteria may be regarded as *predators* — a predator being an organism which kills another organism (its prey) in order to obtain nutrients from it. However, as predators the myxobacteria are atypical: the typical predator ingests its prey before digesting it. In some habitats bacteria themselves are prey for a wide range of protozoan predators, the 'hunters' of the microbial world. Predatory protozoa are often extremely voracious, and one individual may consume many thousands of bacteria during its lifetime; such protozoa may therefore have a profound effect on the numbers of bacteria present in a particular habitat.

A *parasitic* organism obtains its nutrients directly from the cells or body fluids of another *living* organism: the *host*. By living in close physical association with its host the parasite has the benefit of a favourable and stable habitat as well as a ready supply of nutrients; the host may suffer varying degrees of damage — ranging from minor inconvenience to death. Parasitism can be adopted by some free-living bacteria as an alternative mode of life, but in a number of bacteria parasitism is obligatory. Some bacterial parasites can grow only within certain of the host's cells; an example of such an intracellular parasite is *Mycobacterium leprae,* the causal agent of leprosy. Parasites such as these appear to depend heavily on their host's metabolism and often they cannot be cultivated in the laboratory under any conditions other than in living cells. Any parasitic bacterium which affects its host severely enough to cause disease is called a *pathogen;* however, all parasites are not pathogens, and neither are all pathogens parasitic (see eg. *Clostridium botulinum* in Chapter 9).

Bacteria themselves play host to certain specialized parasites. Bacteriophages (Chapter 7) can replicate only within bacterial cells and may therefore be regarded as obligate parasites of bacteria. Bacteria of the genus *Bdellovibrio* are small, motile, vibrio-shaped organisms, widely distributed in soil and water, which are unique in that they parasitize other bacteria — including eg. *Escherichia coli.* A *Bdellovibrio* cell penetrates the cell wall of its host and lodges between the host's cell wall and cell membrane; here it grows and divides, using nutrients from the host cell, until its offspring fill — and eventually lyse — the exhausted host. (Since the host cell is invariably killed, *Bdellovibrio* may equally well be regarded as a predator.)

When a close physical association between two different organisms is of benefit to *both,* the relationship is said to be *symbiotic,* and each partner is called a symbiont. Symbiotic relationships between bacteria and other organisms are quite common in nature. For example, *ruminants* (eg. sheep, goats, cows, deer) are unable to produce enzymes to digest certain components in their diet of plant material. In these animals the alimentary canal includes a specialized compartment (the *rumen*) which contains vast numbers of anaerobic bacteria and other microorganisms. These microorganisms benefit from a warm, stable environment and an abundance of nutrients in the form of vegetation swallowed by the animal; in return they convert the complex and indigestible plant materials to simpler products which can be absorbed and used by the animal.

Perhaps the best-known example of symbiosis in the plant world is that found in leguminous plants such as peas, beans, lupins, clover, etc. The roots of these plants have small swellings or *nodules* which contain bacteria of the genus *Rhizobium*. The bacteria receive both nutrients and protection from the plant, and in return they supply the plant with nitrogen which they 'fix' from the atmosphere (see later). Leguminous plants are thus able to thrive in nitrogen-deficient soils. Similar relationships exist between certain non-leguminous plants and nitrogen-fixing bacteria or cyanobacteria; for example, species of *Alnus* (alder) contain in their roots nitrogen-fixing bacteria of the genus *Frankia*, and the small floating fern *Azolla* contains the nitrogen-fixing cyanobacterium *Anabaena azollae* within special cavities in its leaves.

(We should perhaps mention here that some authors use the term 'symbiosis' to refer to *any* close association between organisms — regardless of whether the organisms derive benefit or harm (or neither) from the association; when used in this sense the term includes not only symbiosis (as described in this chapter) but also eg. parasitism. The use of the term in this way may lead to confusion — which is why we have retained the more familiar and well established meaning of the term.)

Bacteria and the cycles of matter

The elements which make up living organisms occur on Earth in finite amounts; accordingly, if life is to continue the constituent elements of dead organisms must be re-used or 're-cycled', and bacteria (together with other microorganisms) play a vital role in this re-cycling process.

Carbon is the major structural element in living organisms, and the re-cycling of carbon is thus central to the maintenance of life itself. In the biological carbon cycle (Fig. 8-1) the chief bacterial contribution is that of the saprotrophs which degrade dead organic material. Bacteria also make many minor contributions — eg. the fixation of carbon

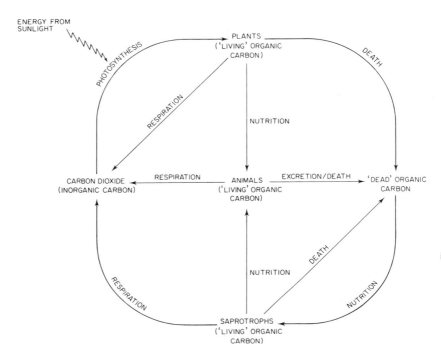

Fig. 8-1 The biological carbon cycle: a simplified cycle showing the major inter-conversions of carbon compounds in nature.

dioxide by autotrophic species. The re-cycling of nitrogen and sulphur is particularly dependent on the activities of bacteria and we shall consider the nitrogen and sulphur cycles in a little more detail.

The nitrogen cycle (Fig. 8-2) Nitrogen is a component of proteins, nucleic acids, amino sugars etc, and is therefore essential to all living organisms. Gaseous nitrogen makes up about 78% of the Earth's atmosphere, but for most organisms gaseous nitrogen is unusable. Many organisms can assimilate inorganic nitrogen in the form of ammonia — primarily by incorporating it in the amino groups of amino acids. Some organisms can also use nitrate as a source of nitrogen, but in this case the nitrate is first reduced to ammonia by *assimilatory nitrate reduction;* this process differs from nitrate respiration (see Chapter 5) in that it does not yield energy, can occur under aerobic or anaerobic conditions, and is carried out only to an extent which fulfills the nitrogen requirement of the cell.

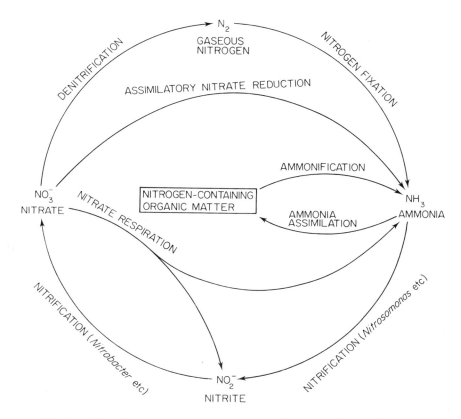

Fig. 8-2   The nitrogen cycle: some interconversions carried out by bacteria.

Where do the ammonia and nitrate come from? The waste products of animals and the dead bodies of all organisms are broken down by a variety of saprotrophs (including bacteria) whose activities include the release of ammonia eg. by the de-amination of amino acids. Some of this ammonia will be used immediately as a source of nitrogen by organisms in the vicinity, but some may be oxidized (by certain chemolithotrophic bacteria) to nitrite — and thence to nitrate — by the process of nitrification (see Chapter 5).

In many organisms nitrate is converted to ammonia and used as a source of nitrogen, as described above. However, some organisms can carry out nitrate respiration (Chapter 5) in which nitrate may be reduced to nitrite, gaseous nitrogen, or ammonia — depending on species and conditions; the production of gaseous nitrogen (*denitrification*) results in the loss of biologically useful nitrogen to the atmosphere.

The supply of biologically usable nitrogen is replenished by the 'fixation' of atmospheric nitrogen: the reduction of gaseous nitrogen to ammonia by certain microorganisms which contain an enzyme complex called *nitrogenase*. Nitrogen fixation requires a large input of energy in the form of ATP and is inhibited in the presence of readily usable forms of nitrogen such as nitrate and ammonia. The fixation of atmospheric nitrogen can be carried out by a number of prokaryotic organisms — including eg. *Azotobacter, Rhizobium,* some photosynthetic bacteria, certain species of *Bacillus* and *Clostridium, Klebsiella pneumoniae,* and the cyanobacteria *Anabaena* and *Nostoc;* some of these are free-living organisms while others occur in symbiotic associations with higher plants. It has become generally accepted in recent years that nitrogen fixation is an exclusive property of prokaryotes; however, this view may need revision in the light of a recent paper (1980) reporting the isolation of a eukaryotic unicellular green alga which is apparently capable of carrying out nitrogen fixation.

Our knowledge of the roles of bacteria in the nitrogen cycle can be put to good use in improving agricultural food production. Food crops are commonly limited in yield by a shortage of available nitrogen in the soil; thus by knowing how nitrogen is lost from the soil — and by exploiting biological nitrogen fixation — we can take appropriate measures to increase crop yields. Nitrogen is taken from the soil when crops are harvested, and may also be lost by the processes of denitrification and nitrification. Denitrification can occur only under anaerobic conditions in the presence of nitrate and organic nutrients — requirements which may be fulfilled eg. in waterlogged soils. The detrimental effects of denitrification are obvious; however, it is not so readily apparent why nitrification leads to the loss of nitrogen, since plants can generally assimilate either nitrate or ammonia. Although both nitrate and ammonia are soluble, ammonium ions adsorb readily to soil particles while nitrate ions do not; thus nitrate is much more readily washed (*leached*) from the soil by rain or flooding.

The adverse activities of bacteria can sometimes be discouraged by eliminating the conditions which favour them. Thus, denitrification can often be reduced by improving soil structure and drainage to minimize the development of anaerobic conditions; additionally, if nitrogenous fertilizers are used they should contain ammonium compounds rather than nitrate. The oxidation of the ammonium compounds to nitrate by nitrification can usually be prevented by adding a nitrification inhibitor to the fertilizer. Although nitrogen lost from the soil can be replaced by the application of nitrogenous

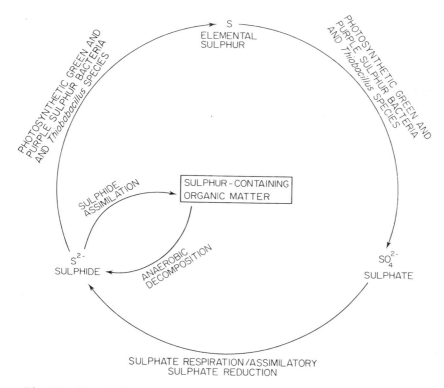

Fig. 8-3 The sulphur cycle: some interconversions carried out by bacteria.

fertilizers, these are expensive and can be afforded least by the countries in greatest need of them. In this context biological nitrogen fixation can be exploited in various ways: for example, by including 'nitrogen-fixing plants' (clover, lucerne etc) in crop rotation schemes, by growing leguminous food crops such as soya beans, by using 'nitrogen-fixing plants' such as *Azolla* as green manure (a practice common in SE Asia), and by encouraging the growth of free-living nitrogen-fixing cyanobacteria in rice paddies. Much recent research has been directed towards the establishment of stable associations between nitrogen-fixing bacteria and non-leguminous crop plants such as wheat; an even more ambitious goal is the transfer of genes coding for nitrogen fixation directly to plant cells with the aim of creating plants capable of carrying out nitrogen fixation without help from prokaryotes.

The sulphur cycle (Fig. 8-3) Sulphur is a component of eg. the amino acids cysteine and methionine, and certain cofactors such as coenzyme A. Green plants, and many bacteria, can assimilate sulphur in the form of sulphate — a nutrient which is usually available in adequate amounts under natural conditions. Prior to its incorporation in cysteine and other thiol-containing compounds, sulphate must be reduced to sulphide by a metabolic process called *assimilatory sulphate reduction;* this process differs from sulphate respiration (Chapter 5) in much the same way as assimilatory nitrate reduction differs from nitrate respiration (see above). Some organisms can also assimilate sulphide directly. In certain types of habitat (eg. in anaerobic aquatic environments) large quantities of sulphide are produced by bacteria which carry out sulphate respiration. This sulphide may in turn be oxidized by chemolithotrophs of the genus *Thiobacillus* and by photosynthetic green and purple sulphur bacteria (see Chapter 5).

9 Bacteria in medicine

Some diseases are caused by a malfunction in the body's chemistry, but many are due to the activities of particular microorganisms on or within the body; any microorganism capable of causing disease is called a *pathogen*. Among diseases of microbial origin some are caused by fungi, some by protozoa, some by viruses, and some — including eg. anthrax, diphtheria, syphilis and tuberculosis — are caused by bacteria. (Many of the diseases caused by bacteria are given in Table 9-2 at the end of this chapter.) In some diseases the link between disease and pathogen is highly specific; such a disease can be caused only by one particular species or strain of bacterium — anthrax, for example, is caused only by *Bacillus anthracis*. In other cases a disease may be due to any of several different causal agents; an example is gas gangrene which can be caused by one (or more) of several different species of *Clostridium*. Sometimes a disease may be caused by an organism which does not usually behave as a pathogen and which may actually be a member of the body's microflora (Table 9-1). For example, species of *Bacteroides* are Gram-negative, anaerobic bacteria which are common in the intestine; nevertheless, they can sometimes cause eg. peritonitis following bowel surgery or accidental damage to the lower intestinal tract. Organisms such as these are called 'opportunist pathogens'.

The routes of infection

The skin is normally an effective barrier to pathogens, but skin may be broken eg. by wounding, bites or surgery. A wound can admit a wide variety of potential pathogens, and those which enter may cause localized infections or systemic diseases (diseases affecting the entire body). Pathogens may also be introduced into the body via insect 'bites'

Table 9-1 The human microflora: some bacteria commonly associated with man.

Skin	*Staphylococcus epidermidis* *Micrococcus* spp *Propionibacterium acnes* *Corynebacterium* spp
Mouth	*Streptococcus salivarius* *Streptococcus mutans* *Bacteroides* spp *Treponema denticola, T.orale,* other treponemes *Actinomyces* spp *Veillonella alcalescens*
Large intestine	*Bacteroides fragilis* *Streptococcus faecalis* *Escherichia coli* *Lactobacillus* spp *Clostridium perfringens* *Proteus* spp
Upper respiratory tract:	
(a) Nostrils	*Staphylococcus epidermidis* *Staphylococcus aureus* *Corynebacterium* spp
(b) Nasopharynx	*Streptococcus* spp *Neisseria* spp *Haemophilus influenzae*
Vagina (adult, pre- menopause)	*Lactobacillus* spp *Corynebacterium* spp *Staphylococcus epidermidis* *Streptococcus* spp *Mycoplasma* spp

— although this method of infection is more common for pathogenic viruses and protozoa than for pathogenic bacteria; bacterial pathogens which may be carried by insects include, for example, the causal agent of bubonic plague, *Yersinia pestis,* which is transmitted from man to man, and from rat to man, by fleas.

The mucous membranes of the intestinal, respiratory, and genito-urinary tracts tend to be more vulnerable than the skin to pathogenic bacteria, and most infections begin in these areas. Thus, in pneumonia and whooping cough, for example, the site of infection is the respiratory tract epithelium, in cholera and typhoid it is the intestinal mucosa, while in urethritis it is the epithelium of the genito-urinary tract.

The mere presence of a pathogen on or within the tissues of a host does not necessarily mean that disease will follow. As we shall see later, the struggle between pathogen and host is by no means one-sided: there are various mechanisms by which the normal healthy body defends itself against microbial attack. Whether or not disease does develop depends on factors such as the *virulence* of the pathogen (ie. its capacity for causing disease) and the host's state of health and his (related) ability to resist disease.

Pathogenesis: the mechanism of disease development

How does a pathogen cause disease? Unfortunately there is no simple answer; different pathogens cause disease in different ways, and only in a few cases do we understand the precise mechanism of disease development.

Bacterial toxins A *toxin* is a substance which is synthesized by a pathogen and which damages or kills the host by interfering with particular host functions. Most pathogenic bacteria form one or more toxins, but in many cases it is not possible to link a toxin directly with a particular symptom of a disease. However, in certain diseases the main symptom(s) are known to be due to the action of one or more specific toxins, and a few of these diseases are discussed briefly below.

(a) *Tetanus* ('lockjaw') involves uncontrollable contractions of the skeletal muscles — often leading to death by asphyxia or exhaustion. The disease develops when deep, anaerobic wounds are contaminated with *Clostridium tetani*. This organism produces a powerful *neurotoxin* — a toxin which acts specifically against nerve tissue. The protein neurotoxin (tetanospasmin) acts mainly on the central nervous system where it blocks the inhibitory action of those cells which normally regulate the nervous stimulation of muscles; this results in the involuntary stimulation (contraction) of skeletal muscles characteristic of the disease.

(b) *Cholera* involves nausea, vomiting, abdominal cramps, and the frequent passage of 'rice-water stools' — ie. watery, mucus-containing stools which may be flecked with blood. The disease is caused by certain strains of *Vibrio cholerae* which, following ingestion, multiply in the lower intestine and liberate a powerful *enterotoxin* — a toxin active against the intestinal mucosa. The toxin (choleragen) acts upon cells of the intestinal epithelium and stimulates the enzyme adenyl cyclase; this in turn raises the intracellular levels of cyclic AMP — thus causing over-secretion of water and electrolytes. Losses of fluids and electrolytes from the body may lead to dehydration and death and are thus major factors in the pathogenesis of cholera.

(c) *Botulism* is a severe and often fatal disease involving muscular paralysis; the symptoms may include weakness, nausea, vomiting, blurred vision, difficulty in swallowing, and a mechanical (muscular) failure of the respiratory system. Symptoms develop following the ingestion of foodstuffs contaminated with one or more of the protein neurotoxins of *Clostridium botulinum*. (Foods commonly implicated in cases of botulism include cold meats and imperfectly canned non-acid vegetables.) The toxins of *C. botulinum* act on the peripheral nervous system; specifically, they bind to nerve-muscle junctions where they inhibit the release of acetylcholine and (hence) inhibit the nervous stimulation of muscles. Botulism is unusual among bacterial diseases in that it results from the ingestion of pre-formed toxin — ie. toxin formed outside the body; ingestion of the living pathogen is not necessary for the production of disease.

(d) *'Food poisoning'* includes those gastro-intestinal diseases caused by the consumption of food containing the enterotoxins of any of several different bacterial pathogens (see Table 9-2); diseases in this category are usually less severe and more common than those discussed above. A particularly common type of food poisoning is caused by certain strains of *Staphylococcus*. In this case symptoms appear some 1-7 hours after the consumption of toxin-contaminated food and involve nausea, vomiting, diarrhoea and prostration; recovery is usually rapid.

Each toxin considered so far is a protein which is released by the pathogen and which has a single, specific mode of action. By contrast, the *endotoxins* of Gram-negative bacteria are lipopolysaccharides which form part of the outer membrane of the pathogen's cell wall (Chapter 2). The presence of endotoxin in the body appears to bring about diverse, non-specific reactions — such as raising body temperature and lowering blood pressure — but the significance of endotoxin in diseases caused by Gram-negative bacteria is still a matter of controversy.

In some cases a toxin can be synthesized only when the cells of the pathogen contain a particular plasmid (Chapter 6) or bacteriophage (Chapter 7). In these cases the genes coding for toxin production are present in the nucleic acid of the plasmid or phage and not in the chromosome of the pathogen; thus, a cell of the same species which lacks the particular plasmid or phage cannot produce toxin. An example of a phage-specified toxin is the diphtheria toxin of *Coryne-bacterium diphtheriae* (see Chapter 7). Some strains of *Escherichia coli* produce a plasmid-specified enterotoxin, and such *enteropathogenic* strains cause intestinal disorders in man and other animals.

Pathogenesis without toxins   In some diseases toxins have little or no known role in pathogenesis. In fact, in a few cases it seems that the mere presence of a pathogen is enough to cause disease — the host actually damaging itself in its attempts to combat infection. Such 'over-reaction' by the body's defence mechanisms seems to be significant in only a few diseases — among them syphilis and tuberculosis; the normal role of the defence mechanisms is to prevent or limit disease rather than to promote it (see later).

Adhesion as a factor in pathogenesis   In a number of diseases (including some of those caused by toxins) pathogenesis appears to involve an early phase in which the pathogen adheres to particular sites on the host. The need for such attachment becomes obvious when we consider, for example, that the common sites of infection, the mucous membranes, are continually flushed by their own secretions and may be subjected to movements such as peristalsis etc — factors which tend to discourage the establishment of a pathogen. Adhesion may also help a pathogen to compete more effectively with the body's normal microflora. The importance of adhesion is well illustrated by the pathogenic role of *Streptococcus mutans:* an organism which occurs in the mouth and which adheres specifically to the teeth. *S. mutans* is a component of dental *plaque* — the film which develops on teeth and at the tooth-gum margins; plaque is composed mainly of bacteria, bacterial products, and salivary substances. Certain products of *S. mutans* are believed to be major factors in the pathogenesis of dental caries (tooth decay).

The body's defence

To any potential pathogen the normal healthy body presents a formidable array of obstacles and barriers. The skin, for example, is more than a simple physical barrier to infection. To most bacteria it is a hostile environment: water is scarce, and the would-be pathogen must also contend with antagonism from the well-adapted skin microflora — some members of which produce antibacterial fatty acids from lipids secreted by the sebaceous glands in the skin. The mucous membranes, too, have their own special mechanisms for defence. The secretions of these membranes actively discourage the establishment of a pathogen both by their mechanical flushing action and by their content of antibacterial substances — eg. the enzyme *lysozyme* (found in tears etc) which disrupts the peptidoglycan of the bacterial cell wall. The mucous surfaces also bear a characteristic microflora with which the pathogen must compete successfully if it is to become established in the body.

If a pathogen succeeds in penetrating the outer barriers of the body it is immediately faced with the inner defences. Within the tissues and circulatory systems there are certain specialized cells called *phagocytes* which engulf and destroy particles of 'foreign' matter — including microorganisms; these scavenging cells can usually prevent the establishment of a pathogen in the nutrient-rich environment of the inner tissues. There are, however, some pathogens which can fight back by secreting substances (called *leucocidins*) which kill phagocytes; pathogens which form leucocidins include certain strains of *Staphylococcus* and *Streptococcus*. Another way of eluding the phagocytes has evolved in certain strains of *Streptococcus* whose capsules (Chapter 2) contain hyaluronic acid; this substance is also a normal constituent of animal tissues so that the capsule provides the pathogen with a form of 'camouflage' which appears to give it some protection against the phagocytes.

Should the pathogen persist at a particular site in the tissues it may provoke a so-called *inflammatory response*. Inflammation is characterized by reddening, swelling, heat, and pain at the affected site. It is a non-specific response which can be caused by any of a variety of agents — including heat and chemical irritants as well as microbial pathogens; nevertheless, some of the effects of inflammation can be inhibitory to a pathogen. For example, one feature of inflammation is the increased outflow of fluid (plasma) from the local blood capillaries to the surrounding tissues — which therefore become swollen; since plasma normally contains antimicrobial factors, this helps to make the inflammatory response an important part of the body's generalized reaction to pathogens.

So far we have considered only *constitutive* defences: non-specific defence mechanisms which are normally faced by every bacterial pathogen attempting invasion. There are also more specific types of defence which the body can bring into play only after contact with a pathogen; a response of this type is said to be *inducible,* and such a response can generally act against only that pathogen whose presence induced it. This means that the body can learn to recognize — and fight against — a particular pathogen, but how? Parts of a bacterial cell (eg. components of its cell wall) may act as *antigens:* substances whose presence within the body stimulates certain white blood cells to synthesize *antibodies;* an antibody is a protein which can combine *specifically* with the antigen that induced its synthesis. Since different pathogens tend to be chemically dissimilar — ie. tend to have different antigens — antibodies induced by a given pathogen will generally combine only with that pathogen. How then does this help the body to fight that pathogen? When an antibody combines with an antigen the antigen-antibody complex automatically binds to certain protein factors which occur in normal plasma — factors which form part of a system of plasma proteins known as *complement.* By binding certain components of complement, the antigen-antibody complex becomes 'sticky' — ie. it adheres strongly to certain phagocytic cells (*immune adherence*). Hence, by combining with the cell surface antigens of pathogenic bacteria, antibodies increase the susceptibility of those bacteria to phagocytosis. (Even without complement, antibodies can increase the susceptibility of cells to phagocytosis simply by combining with their cell-surface antigens — a process known as *opsonization.*) Furthermore, when antibodies combine with certain Gram-negative bacteria in the presence of complement the bacteria are lysed — a phenomenon called *immune lysis.*

Just as various components of a pathogen can act as antigens, so too can many bacterial products — including toxins. The combination of a toxin with specific antibody abolishes the harmful properties of that toxin and assists in its elimination from the body.

The production of antibodies can be stimulated artificially by *vaccination* — the object of which is to help the body defend itself in the event of a subsequent attack by a given pathogen. In vaccination antigens of the pathogen (in the form of killed cells or cell components) are introduced into the body by injection or, sometimes, by oral administration. The body responds by producing the corresponding anti-pathogen antibodies; additionally, the specific antibody-producing cells become 'primed' — ie. able to respond, quickly, with increased production of antibody on a subsequent exposure to the pathogen. It is also possible to vaccinate against the effects of a toxin (such as the tetanus toxin) by administering an inactivated form of the toxin (a *toxoid*) — thus inducing the formation of anti-toxin antibodies.

Another type of inducible defence mechanism is mediated by cells rather than by antibodies. In *cell-mediated immunity* certain white blood cells (the T lymphocytes) react to specific antigens eg. by liberating chemicals called *lymphokines* which — among other things — attract certain phagocytes (the macrophages) and enhance their phagocytic activities. Cell-mediated immunity is believed to be an important part of the body's defence against a range of microbial pathogens, including some pathogenic bacteria. However, in some diseases the reactions of cell-mediated immunity appear to contribute to the symptoms; in tuberculosis, for example, at least some of the tissue damage which occurs at the local foci of infection (tubercles) appears to be associated with the concentration of white blood cells at these sites.

The spread of disease within a population

Some pathogens are commonly associated with particular parts of the environment: *Clostridium tetani,* for example, occurs in many types of soil, while the strains of *Leptospira* that cause infectious jaundice (leptospirosis) are quite common in water contaminated with rats' urine. In general, the diseases caused by such pathogens do not usually spread between individuals in a population — often because there is no way in which the pathogen can pass easily from one person (or animal) to another.

Many diseases can spread from person to person either directly or indirectly. In direct spread the pathogen is transmitted by physical contact between an infected person and a healthy one. In only a few diseases is it usual for spread to occur in this way, and these diseases are generally caused by pathogens which are unable to survive for long outside the body; an example is the venereal disease syphilis, caused by the spirochaete *Treponema pallidum.*

Most diseases spread *indirectly* from person to person — the pathogen being transmitted in a way which is usually related to its normal route of infection. For example, pathogens which infect via the intestine are commonly transmitted in contaminated food or water, and diseases such as gastro-enteritis, dysentery, typhoid and cholera are usually spread in this way. The contamination of food or water with a pathogen of this type generally involves some connection between the food or water and the faeces of a patient who is suffering from the disease; contamination may occur, for example, when the hands of a food-handler carry traces of faecal matter, when a housefly lands alternately on faeces and food, or when sewage has leaked into a source of drinking water. Usually the pathogen can be traced back — via the food or water — to an individual who is suffering from the disease. Occasionally, however, the source of the pathogen is a person who is *not* suffering from the disease but who is nevertheless playing host to the pathogen and acting as a reservoir of infection; such an apparently healthy source of pathogenic organisms is known as a *carrier*. One notorious carrier, a domestic cook by the name of Mary Mallon, transmitted typhoid fever to nearly thirty people before she was traced, and the name 'typhoid Mary' is sometimes used to refer to an actual or suspected carrier in outbreaks of typhoid and other diseases.

Pathogens which infect via the respiratory tract are often transmitted from person to person by so-called *droplet infection*. When a person coughs or sneezes, or even speaks loudly, minute droplets of saliva are expelled from the mouth; these droplets may carry with them pathogens which may have been present on the surface of the respiratory tract. Since the smaller droplets can remain in the air for some time, they can be inhaled by other individuals and can therefore act as vehicles for the transmission of pathogens. Diphtheria, whooping cough, and tuberculosis are examples of diseases which are transmitted in this way.

A few bacterial pathogens are transmitted from one person to another by a third organism called a *vector*. For example, bubonic plague (caused by *Yersinia pestis*) is transmitted by fleas, while epidemic typhus (caused by *Rickettsia prowazekii*) is typically transmitted by lice.

The prevention and control of transmissible diseases

Once we know how a disease can spread we can often devise methods for preventing or limiting its spread. Clearly, those diseases which spread only by direct physical contact can be prevented simply by avoiding such contact. For other transmissible diseases prevention or control may involve blocking the pathway which the pathogen normally takes from one person to another. Thus the spread of diseases such as typhoid and cholera can be halted by measures which may include: (i) improvement in personal hygiene — eg. washing hands after visiting the lavatory; (ii) protection of food etc. from flies and other insects likely to carry pathogens, and reduction of the numbers of such insects by the use of insecticides; (iii) protection of drinking water from contamination by sewage, and the effective treatment of communal water suplies with antimicrobial agents such as chlorine; (iv) disinfection of small quantities of untreated water before consumption — eg. by boiling or by treatment with a disinfecting agent such as halazone.

Diseases which spread by droplet infection are generally more difficult to deal with. The physical exclusion of droplets (eg. by face masks) is usually not practicable, and one of the principal control measures in diseases such as diphtheria and whooping cough involves the protection of susceptible individuals by vaccination. In such diseases there is also some advantage in isolating sick individuals until they are no longer able to transmit the pathogen.

A disease which is spread by a vector can be effectively controlled by eliminating the vector or reducing its numbers. Such control is applicable in the case of diseases such as typhus and bubonic plague.

Table 9-2 Some diseases caused by bacteria.

Disease	Causal agent	Main symptoms of disease	Route of infection
Anthrax	*Bacillus anthracis*	Localized pustule (anthrax boil) or lung infection (Woolsorters' disease). May lead to septicaemia.	Via wounds or by inhalation.
Botulism	*Clostridium botulinum*	Weakness, vomiting etc. with eventual paralysis of respiratory muscles.	Ingestion of toxin.
Cholera	*Vibrio cholerae* (some strains)	Vomiting, muscular cramps, passage of watery stools leading to dehydration.	Ingestion of pathogen.
Diphtheria	*Corynebacterium diphtheriae* (toxin-producing strains)	Membrane forms on tonsils and/or on pharyngeal wall — may inhibit breathing; also systemic symptoms due to toxin.	Inhalation or ingestion of pathogen.
Dysentery	eg. *Shigella* species	Abdominal pain, fever, diarrhoea with passage of blood and pus.	Ingestion of pathogen.
Erysipelas	*Streptococcus pyogenes*	Localized inflammation of skin; may lead to septicaemia.	Via wounds etc.
"Food poisoning"	eg. *Staphylococcus* spp., *Bacillus cereus, Clostridium perfringens, Salmonella typhimurium*	Generally involves nausea, vomiting and diarrhoea.	Ingestion of toxin and/ or pathogen.
Gas gangrene	eg. *Clostridium perfringens* type A, *C.nozyi, C.septicum*	Local but spreading death and degeneration of tissues around infected site; pockets of gas form in tissues.	Via wound or other lesion (eg. boils).

Table 9-2 (*continued*)

Disease	Organism	Symptoms	Transmission
Gonorrhoea	*Neisseria gonorrhoeae*	Inflammation of genito-urinary tract; other symptoms may also occur.	Via sexual contact.
Legionnaires' disease	*Legionella pneumophila*	Fever, coughing, malaise, pneumonia.	Probably by inhalation of pathogen.
Leprosy	*Mycobacterium leprae*	Skin lesions, involvement of peripheral nerves leading to weakness and anaesthesia.	Probably via wounds or via mucous membranes.
Meningitis	eg. *Neisseria meningitidis*, *Haemophilus influenzae*, *Streptococcus pneumoniae*, *Listeria monocytogenes*	Inflammation of the meninges (membranes surrounding brain and spinal cord).	Via wounds or sinuses.
Plague (bubonic)	*Yersinia pestis*	Infected, swollen, necrotic lymph nodes (buboes). (Other forms of the disease occur.)	Via flea bites.
Pneumonia	eg. *Streptococcus pneumoniae*, *Haemophilus influenzae*, *Klebsiella pneumoniae*, *Mycoplasma pneumoniae*.	Inflammation of the lungs (often following weakening illness such as virus infection of the respiratory system).	Inhalation of pathogen.
Q fever	*Coxiella burnetii*	Fever, malaise, headache, muscular pain.	Inhalation of contaminated dust, ingestion of contaminated milk.

Disease	Pathogen	Symptoms	Transmission
Scarlet fever	*Streptococcus pyogenes* (toxin-producing strains).	Sore throat, fever, rash.	Via mouth or nose.
Syphilis	*Treponema pallidum*	Ulcerative lesions on mucous membranes and/or skin – may be followed by lesions in the heart, central nervous system, etc.	By direct contact – especially sexual contact.
Tetanus (lockjaw)	*Clostridium tetani*	Sustained and involuntary contractions of skeletal muscles.	Via wounds or other lesions.
Trachoma	*Chlamydia trachomatis*	Inflammation of conjunctivae; may lead to partial or total blindness.	Contamination of conjunctivae by pathogen.
Tuberculosis, pulmonary (consumption)	*Mycobacterium tuberculosis*, *Mycobacterium bovis*	Lesions (tubercles) develop in the lungs. (Other forms of the disease occur.)	Inhalation of pathogen.
Typhoid	*Salmonella typhi*	Fever, lesions on skin and in intestinal and lymphoid tissue; delirium or stupor may occur.	Ingestion of pathogen.
Typhus (epidemic or louse-borne typhus)	*Rickettsia prowazekii*	Prostration, muscular pains, sustained fever; rash on trunk and limbs. Delirium or stupor may occur.	Contamination of bite/wound with louse faeces containing pathogen.
Whooping cough	*Bordetella pertussis*	Paroxysms of coughing, each followed by an audible inspiratory "whoop".	Inhalation of pathogen.

10 Man against bacteria

Bacteria can be a nuisance — or even dangerous — in many everyday situations, and we therefore need methods by which we can eliminate them or inhibit their activities. Of the many methods available the one chosen in any given situation will depend largely on the nature of the situation. Sometimes we need to destroy, completely, all forms of microbial life on a given object — as, for example, when surgical instruments are prepared for use in an operation. At other times less rigorous measures may be adequate, and it may be sufficient to eliminate only the potentially harmful organisms. There is also the special problem of inactivating pathogenic bacteria within the living tissues of an infected person. In this chapter we shall consider, in turn, each of these aspects of our war against bacteria.

Sterilization

A 'total kill' involving the destruction of *all* living organisms — including viruses and bacterial endospores — requires fairly drastic measures and is called *sterilization*. The ideal method of sterilization should be quick, simple, efficient, and applicable to a wide range of materials; in practice such requirements are most easily met by physical methods — especially by the use of heat.

Sterilization by heat Bacteria vary in their susceptibility to heat, and endospores are much more resistant than are vegetative cells; vegetative cells are generally killed rapidly by boiling water, while endospores may survive such treatment for considerable periods of time. The sterilizing power of heat depends not only on the temperature used but also on factors such as time, the presence of moisture, and the number and condition of the microorganisms present.

116

The use of *fire* is an extreme form of heat sterilization. We have already discussed the technique of flaming for the rapid sterilization of loops etc (Chapter 4); similarly, disposable items such as used dressings and used disposable syringes may be sterilized — and destroyed — by incineration. However, such methods are obviously of limited application, and less destructive methods are necessary for most purposes.

The *hot-air oven* is widely used for the sterilization of heat-resistant items such as clean glass-ware. Items placed in the oven are maintained at a temperature of 160-170°C for about 1-1½ hours. Since air is not a good conductor of heat the hot air should be circulated by a fan to ensure that all parts of the oven are kept at the required temperature; items within the oven should therefore be well spaced in order not to impede air circulation.

Moist heat is more effective than dry heat as a sterilizing agent, and *steam* can sterilize at lower temperatures (or shorter times) than those used in the hot-air oven. However, steam at normal atmospheric pressure (760mm mercury) has a temperature of 100°C and, as we have seen, endospores may be able to survive for some time at this temperature. Nevertheless, steam can reach higher temperatures — suitable for sterilization — when its pressure is increased; in fact, there is a definite relationship between the pressure and temperature of pure steam: the higher the pressure the higher the temperature. Thus, when sterilizing by steam it is necessary to use a special vessel (an *autoclave*) which can contain the steam under pressure. Items to be sterilized are placed inside the chamber of the autoclave which is then closed; steam is introduced into the chamber, displacing the air, and its pressure is allowed to build up to a level determined by the setting of an adjustable valve. The setting of this valve thus determines the temperature within the autoclave. The time allowed for sterilization must be sufficient for all parts of the items to reach the sterilizing temperature and to stay at that temperature until any organisms present have been killed. Temperature/time combinations which are commonly used when autoclaving are 115°C/35 minutes, 121°C/15-20 minutes, and 134°C/4 minutes.

For effective sterilization the quality of the steam used in an autoclave is important. It must be saturated — ie. it must hold as much water in vapour form as is possible at the given temperature and pressure. Also, there must be no air present in the steam during the sterilization process because air upsets the pressure-temperature relationship of pure steam: an air-steam mixture at a given pressure

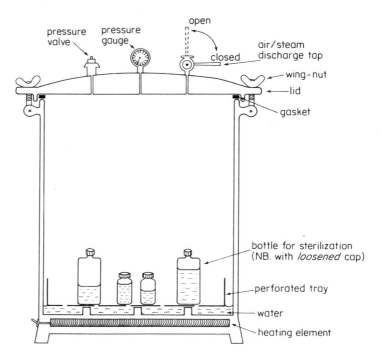

Fig. 10-1 A typical small, portable laboratory autoclave. Mode of operation: (a) Water is placed in the bottom of the autoclave chamber. (b) Objects for sterilization are placed on the perforated tray which holds them clear of the water. (c) The lid — with the air/steam discharge tap *open* — is clamped securely in position; the rubber gasket ensures an air-tight seal. (d) The heating element is switched on to heat the water in the bottom of the chamber. Steam fills the chamber and eventually displaces the air through the discharge tap. (e) Pure steam begins to issue vigorously from the discharge tap, which is then closed. As heating is continued, water continues to vaporize so that there is an increase in pressure — and hence temperature — within the chamber. (f) Once the desired pressure/temperature has been reached, the (pre-set) valve opens to allow steam to escape, thus maintaining the pressure at a steady level. (g) When the appropriate time has elapsed (see text) the heating element is switched off and the autoclave is allowed to cool down until the pressure inside the chamber (indicated by the gauge) does not exceed atmospheric pressure; the lid can then be opened safely and the sterile contents removed.

has a lower temperature than that of pure steam at the same pressure. Hence, air must be completely purged from the chamber — and from all items within the chamber — before the pressure is allowed to rise, otherwise the valve will open before the sterilizing temperature is reached.

Small portable laboratory autoclaves generally resemble the ordinary domestic pressure cooker both in principle and in mode of use

(Fig. 10-1). In larger autoclaves — such as those used in hospitals and industry — steam is usually piped to the autoclave chamber from a boiler, and factors such as timing, pressure and steam quality are often controlled automatically. In some models steam is admitted at the top of the chamber so that air is displaced downwards; this is more effective than upward displacement (as used in smaller autoclaves) because steam is lighter than air under these conditions. In another type of autoclave air is removed from the chamber by a vacuum pump before steam is admitted; this allows rapid and thorough penetration by the steam of porous materials such as dressings or bed-linen — materials which tend to trap air.

Some materials cannot be sterilized by autoclaving; these include water-repellent substances (eg. petroleum jelly) and substances which are volatile or heat-labile (ie. destroyed by heat). Some of these materials (such as petroleum jelly) can be sterilized in a hot-air oven. Materials which are damaged by normal autoclave temperatures may be sterilized by steam at reduced pressure (to give a temperature of eg. 80°C) used in conjunction with a small amount of formaldehyde. This method kills endospores within about 2 hours and is used for sterilizing heat-sensitive surgical instruments, plastic tubing, woollen blankets, etc. (If the formaldehyde is omitted, vegetative cells will be killed but endospores will usually survive; in this case the process cannot be called sterilization.)

Sterilization by ionizing radiation Ionizing radiations may be effective sterilizing agents under suitable conditions, but their use requires expensive, bulky equipment which must be operated by skilled personnel. *Gamma* radiation is widely used commercially for sterilizing pre-packed disposable items such as plastic petri-dishes and syringes.

Sterilization by filtration A liquid which cannot be autoclaved may be sterilized by passing it through a filter with very small pores. A filter with pores of 0.30-0.45μm will retain most bacteria, but one with smaller pores is necessary if the smallest bacteria and viruses are to be retained. During filtration the liquid may be drawn through the filter by reduced pressure in the receiving vessel, or it may be forced through eg. by a syringe plunger. The filter itself commonly consists of a thin membrane of cellulose acetate or similar material. Filtration is used for sterilizing serum (for laboratory use), solutions of heat-labile antibiotics, media containing heat-labile sugars, etc.

Sterilization by chemical agents Chemicals used for sterilization (*sterilants*) are of necessity highly reactive and damaging to living tissues; they therefore require careful handling, and tend to be used only in larger institutions which can provide suitable equipment and personnel. One of the more commonly used sterilants is *ethylene oxide* (C_2H_4O) — a water-soluble cyclic ether. Ethylene oxide is a gas at temperatures above 10.8°C and forms explosive mixtures with air; it is therefore used diluted with another gas such as carbon dioxide or nitrogen. When used for sterilization the gas mixture is contained within a special chamber, and the conditions of temperature, humidity, time, and concentration of ethylene oxide must be carefully controlled. Ethylene oxide can be used for sterilizing clean medical equipment, bed-linen, and heat-labile materials such as certain plastics.

Disinfection

Measures which kill or inactivate potentially harmful microorganisms — but which do not necessarily kill *all* the microorganisms present — come within the category of *disinfection*. Disinfection commonly has little or no effect on bacterial endospores. Although chemicals are widely used for disinfection, there are occasions when physical methods may be more suitable.

Disinfection by pasteurization Pasteurization is a mild heat treatment named after the great pioneering microbiologist Louis Pasteur. Pasteur discovered that if wine was heated to temperatures of 50-60°C, contaminating bacteria were killed and spoilage of wine was prevented. Pasteurization is used today for treating milk and other dairy products, vinegar, some wines and beers, and various other products — the aim being to destroy any pathogens which may be present and/or to enhance the keeping qualities of the product by destroying vegetative spoilage organisms. The temperature/time combinations used for pasteurization vary. Milk, for example, is usually heated to at least 72°C and held at this temperature for at least 15 seconds — a treatment which may kill 90% or more of the vegetative bacteria present; however bacterial endospores and some (*thermoduric*) vegetative bacteria survive.

Disinfection by ultraviolet radiation Ultraviolet rays damage DNA and can be lethal to bacteria under appropriate conditions. However, the rays cannot penetrate solids or liquids to any significant degree, and their use in disinfection is restricted to reducing the numbers of bacteria present in the air and on surfaces — eg. in operating theatres and food-handling rooms.

Disinfection by chemicals Chemicals used for disinfection may be divided into two categories: *disinfectants,* which are used for treating inanimate objects, and *antiseptics,* which can be applied safely to skin and other tissues. The general properties of disinfectants, as described below, apply equally to antiseptics.

Any given disinfectant is usually more effective against some bacteria than others, and disinfectants generally have little or no effect on endospores. The activity of a particular disinfectant may be affected quite drastically by factors such as dilution, temperature, pH, and the presence of hard water, organic matter, soaps or detergents; additionally, a disinfectant can be effective only when allowed to act for an appropriate period of time. Some disinfectants are unstable when diluted and lose activity if stored in this state; certain bacteria — particularly species of *Pseudomonas* — can actually grow in some diluted disinfectants.

Some disinfectants can *kill* vegetative bacteria, and these are said to be *bactericidal.* Others merely halt the growth of bacteria, and if they are inactivated — eg. by dilution or by chemical neutralization — the bacteria may be able to resume growth; such disinfectants are said to be *bacteriostatic.* A bactericidal disinfectant may become bacteriostatic on dilution.

Of the disinfectants and antiseptics currently in use we can afford space to mention only a few. *Phenolic* disinfectants include substituted phenols (eg. cresols, xylenols) and compounds containing two phenolic groups (bisphenols). Examples include 'Lysol', a disinfectant which contains a mixture of cresols solubilized by soap; 'Dettol', an antiseptic and disinfectant which contains chloroxylenols; 'Hexachlorophene', a bisphenol which is used as a skin antiseptic. *Halogens* and many of their derivatives are effective disinfecting agents; chlorine is widely used for the disinfection of water supplies, while iodine is an effective antiseptic. Sodium hypochlorite (household bleach) is a good disinfectant widely used in the dairy and food industries. *Alcohols* can be bactericidal under appropriate conditions, and a mixture of 70% ethanol in water is used as a skin antiseptic. *Soaps* have little or no antibacterial activity unless they contain added antiseptics — such as phenol in carbolic soap; however, the use of soap for skin cleansing may be beneficial in that bacteria are removed from the skin along with dirt and grease. *Quaternary ammonium compounds* are cationic detergents which are used as antiseptics and as disinfectants for dairy equipment etc; they include eg. cetyltrimethylammonium bromide — marketed as 'Cetrimide', 'Cetavlon', etc. These compounds are inhibited by low pH, by certain types of organic compounds, and by soaps, anionic detergents, and some metal ions; they are more effective against Gram-positive than Gram-negative bacteria.

Antibiotics

Originally the term *antibiotic* referred to any substance which is produced by a microorganism and which, even at very low concentrations, inhibits or kills certain other microorganisms; the term is now generally used in a wider sense to include, in addition, any semi-synthetic or wholly synthetic substance with these properties.

No antibiotic is effective against all bacteria. Some are active against a narrow range of species, while others are active against a broad spectrum which includes both Gram-positive and Gram-negative species. In some cases a natural (microbially-produced) antibiotic can be chemically modified in the laboratory to give a 'semi-synthetic' antibiotic which may have a significantly different spectrum of activity. Like disinfectants, antibiotics are either bactericidal or bacteriostatic in their action, and an antibiotic which is bactericidal at one concentration may be bacteriostatic at a lower concentration.

An antibiotic characteristically acts at a precise site in a bacterial cell; for example, depending on antibiotic, the site of action may be the cell wall, the cell membrane, the protein-synthesizing machinery, or an enzyme involved in nucleic acid synthesis. Since a bacterial cell differs in many ways from an animal cell, the toxic effect which an antibiotic has on a bacterium is unlikely to be exerted on an animal cell — ie. antibiotics are usually *selective* in their action. Some antibiotics can therefore be used to treat diseases caused by pathogenic bacteria — the antibiotic selectively attacking the pathogen within the body. Clearly, any antibiotic used in this way must be generally non-toxic to animal tissues and must remain active within the body for long enough to be effective against the pathogen. Of the vast range of known antibiotics only a relatively small number are suitable for use in treating disease, and a few of these are mentioned briefly below.

Penicillins (produced by fungi of the genus *Penicillium*) disrupt the synthesis of peptidoglycan in the cell walls of growing bacteria.

Penicillin G (one of the first antibiotics to be used clinically) is active mainly against Gram-positive bacteria but other, chemically modified, penicillins — such as ampicillin — have increased activity against Gram-negative species. *Cephalosporins* (produced by fungi of the genus *Cephalosporium*) also disrupt peptidoglycan synthesis. *Polymyxins* (produced by *Bacillus polymyxa*) increase the permeability of the bacterial cell membrane and are effective mainly against Gram-negative bacteria. *Chloramphenicol* is a broad-spectrum antibiotic originally obtained from *Streptomyces venezuelae* but now synthesized chemically; it binds to the 50S subunit of bacterial ribosomes and prevents the formation of peptide bonds during protein synthesis. The aminoglycoside group of antibiotics includes eg. *gentamicin, kanamycin, neomycin* and *streptomycin* — all broad-spectrum antibiotics produced by actinomycetes. The aminoglycosides bind to the 30S subunit of the bacterial ribosome, thereby disrupting protein synthesis. The *tetracyclines* (produced by species of *Streptomyces*) are also broad-spectrum antibiotics which bind to 30S ribosomal subunits; they prevent amino acyl-tRNA molecules from binding to the A site on the ribosome (see Fig. 6-9). *Novobiocin* (obtained from *Streptomyces* species) specifically inhibits bacterial DNA polymerase, thus preventing DNA replication; it is more active against Gram-positive than Gram-negative bacteria. *Nalidixic acid* (a synthetic antibiotic) inhibits the replication of bacterial DNA and is active against a range of Gram-negative species. The *rifamycins* (from the actinomycete *Nocardia mediterranei*) specifically inhibit bacterial RNA polymerase, preventing the initiation of transcription; Gram-positive species tend to be more sensitive than Gram-negative species to the rifamycins. The *sulphonamides* are synthetic compounds which interfere with the synthesis of folic acid — a coenzyme essential in a number of vital metabolic reactions. A sulphonamide molecule is similar in shape to a normal component of folic acid, *p*-aminobenzoic acid (PABA); during folic acid synthesis, sulphonamide may be incorporated instead of PABA — resulting in the formation of an inactive analogue of folic acid.

Why are some bacteria not affected by some antibiotics? In some cases a bacterium is resistant to a particular antibiotic because it lacks the target structure affected by that antibiotic; for example, species of *Mycoplasma* (which lack cell walls) will not be affected by penicillins whose target is a cell wall component. Some bacteria may not carry out the particular process affected by an antibiotic; thus, for example, sulphonamides will not be effective against organisms which normally obtain their folic acid, ready-made, from the environment. Resistance can also be due to the ability of a cell to exclude an antibiotic from the target site; in many Gram-negative bacteria the outer membrane is impermeable to certain antibiotics, and in both Gram-positive and Gram-negative species the cell membrane may also act as a barrier. Some bacteria can produce enzyme(s) which inactivate particular antibiotic(s); certain strains of *Staphylococcus,* for example, produce penicillinase — an enzyme which can inactivate many of the penicillins.

Resistance to a particular antibiotic can also be *acquired* by cells which were previously sensitive to that antibiotic. This may happen eg. as a result of a mutation or by the acquisition of an R plasmid (Chapter 6). A mutation may alter the target site in the cell in such a way that it is no longer affected by the antibiotic; for example, a mutation which confers resistance to streptomycin may alter the target site (the 30S ribosomal subunit) in such a way that either the antibiotic can no longer bind to it or, if binding does occur, it no longer affects ribosomal function. A single mutation usually confers resistance to only one antibiotic — or to closely related antibiotics which have the same site of action. In contrast, an R plasmid may carry genes for resistance to one or to many related or unrelated antibiotics. Genes carried by R plasmids may code for enzymes which inactivate antibiotics (in some cases the gene for penicillinase is plasmid-borne) or they may bring about changes in the permeability of the cell to particular antibiotics.

11 The identification of bacteria

Given an unknown species of bacterium how does the bacteriologist set about identifying it? Usually he determines its characteristics and compares them with those of each of a number of known named species until a 'match' is found. The principle is simple enough, but in practice which characteristics does he determine — and must the unknown species be compared with each of the thousands of known species of bacteria? Fortunately, the source of an unknown species often gives certain clues which — together with a few simple observations and tests — may indicate the possible identity of the organism or, at least, serve to narrow the search to one or other of the major groups of bacteria. For example, if the bacterium is known to have been isolated from soil, and is found to be a Gram-positive, endospore-forming bacillus, the bacteriologist would think immediately of the family Bacillaceae (Chapter 12) since this family contains all the Gram-positive endospore-forming bacilli — most of which occur in soil. Identification to species level is then carried out by comparing the characteristics of the unknown species with those of genera and species in the family Bacillaceae; characteristics which distinguish members of other families need not be considered. Clearly, the practice of identification is made easier if the bacteriologist (i) has a knowledge of the types of organism likely to be present in a given situation, and (ii) is familiar with the main distinguishing features of the major groups of bacteria.

In most cases the first step in identifying a bacterium is to obtain it in pure culture (Chapter 4). The reason for this is obvious. If an identification test is carried out with an impure culture (a mixture of organisms) the effects of one organism may conflict with those of another so that generally it will be impossible to obtain a meaningful result. However, even with a pure culture it is occasionally found that the characteristics of the unknown strain of bacterium do not match, exactly, those of any species given in a manual of identification. This can occur, for example, when the unknown organism is a mutant strain (Chapter 6) in which the mutation has altered one or more of the organism's 'typical' characteristics; similarly, a strain which contains a plasmid (Chapter 6) may display characteristics which are unlike those of the species to which the strain belongs. Nevertheless, the organism is not always to blame: 'inexplicable' results can sometimes be due to slight variations in the methods and/or materials used in the tests themselves.

In identifying an unknown bacterium (obtained in pure culture) the following characteristics are often the first to be determined since they have the greatest *differential* value — ie. each characteristic helps to exclude one or more of the major groups of bacteria: (i) reaction to certain stains, particularly the Gram stain; (ii) morphology of the cells (coccus, bacillus etc); (iii) motility; (iv) the ability to form endospores; (v) the ability to grow under aerobic and/or anaerobic conditions; (vi) the ability to produce the enzyme catalase. Fortunately these characteristics are among the easiest to determine.

Preliminary observations and tests

Reaction to stains The reaction of bacteria to a stain can be determined by applying the stain to a thin film of cells (a *smear*) prepared on a glass slide. A loopful of distilled water is placed on a clean glass microscope slide. With a loop, a speck of growth from a colony is transferred to the water and mixed with it ('emulsified') to form a slightly milky suspension of cells. Using the loop, the suspension is spread over an area of one or two square centimetres and allowed to dry — forming the smear; the smear is 'fixed' by passing it quickly through the bunsen flame once or twice and is then ready for staining and subsequent examination under the microscope.

The Gram stain The background to this important staining procedure has been given in Chapter 2. One of many versions of the procedure is as follows. A heat-fixed smear (prepared as above) is stained for one minute with an alcoholic solution of crystal violet; the stained smear is briefly rinsed with Lugol's iodine (a solution of iodine and potassium iodide in water) and is then flooded with Lugol's iodine for a further one minute. Decolorization is then attempted by treating the smear with 95% ethanol. Using a pasteur pipette the ethanol is allowed to run over the surface of the smear while the slide is held in a tilted position. Decolorization is continued only for as long as the stain runs freely from the smear (a few seconds); the smear is then rinsed immediately under a running tap. At this stage any Gram-positive cells in the smear will be stained violet while any Gram-negative cells will be unstained. The smear is now *counterstained* — the purpose of which is to ensure that, under the microscope, Gram-negative cells (if present) will be clearly visible and distinguishable from Gram-positive cells. The smear is counterstained with eg. dilute carbolfuchsin for one minute, rinsed briefly under the tap, and blotted dry. When viewed under the microscope Gram-positive cells (violet) should be clearly distinguishable from Gram-negative cells (red).

Certain species of bacteria do not give a consistent reaction to the Gram stain — sometimes reacting positively, sometimes negatively; these bacteria are said to be *Gram-variable*.

The acid-fast stain (Ziehl-Neelsen's stain) 'Acid-fast' bacteria differ from all other bacteria in that once they are stained with hot concentrated carbolfuchsin they cannot be decolorized by mineral acids or by mixtures of acid and ethanol; the acid-fast bacteria include species of *Mycobacterium* and certain other members of the order Actinomycetales. Acid-fastness is easily demonstrated as follows. A heat-fixed smear is flooded with a concentrated solution of carbolfuchsin and the slide heated until steam appears; the stain should not boil. The slide is kept hot for about five minutes, left to cool, and rinsed in running water. Decolorization is attempted by passing the slide through several changes of acid-alcohol (eg. 3% concentrated hydrochloric acid in 90% ethanol). After washing in water, the smear is counterstained with a contrasting stain such as malachite green, washed again, and dried. Acid-fast cells stain red, other cells stain green.

Morphology Morphology is generally determined by examining a stained smear under the microscope. The smear can be stained either by Gram's method or by a simpler procedure — for example, by treating a heat-fixed smear for one minute with a solution of methylene blue in a mixture of ethanol and water. Usually the stained smear is examined under the oil-immersion lens of the microscope (total magnification about $1000\times$) although the cells of some species (eg. *Bacillus megaterium*) are clearly visible under the 'high-dry' lens (total magnification about $400\times$). In some cases morphology can be determined by examining a suspension of unstained cells under the phase-contrast microscope.

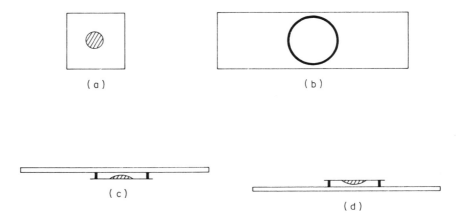

Fig. 11-1 The hanging-drop method for determining motility. (a) A drop of culture containing live, unstained bacteria is placed on a clean cover-slip. (b) A ring of plasticine is pressed onto a microscope slide. (c) The slide is inverted and pressed onto the cover-slip. (d) The whole is inverted and examined under the high-power objective lens of the microscope.

Motility Motility (Chapter 2) is a feature of many species of bacteria; some major groups (such as the family Bacillaceae) include both motile and non-motile species. Flagellar motility is most easily observed by examining a 'hanging-drop' preparation (Fig. 11-1) under the high power of the microscope. With the light source suitably adjusted the cells, although unstained, should be sufficiently visible to show whether or not they are motile — ie. whether or not the cells change positions relative to one another. (Cells can be seen more easily by using dark-field or phase-contrast microscopy.) Motility should be distinguished from Brownian motion. The latter term refers to small random movements exhibited by any small particle (of about 1μm or less) when freely suspended in a fluid medium; these movements are due to bombardment of the particles by molecules of the medium. (Compare Brownian motion with the motility of *Escherichia coli* described in Chapter 2.)

The motility of anaerobes (eg. species of *Clostridium*) cannot be determined by the hanging-drop method. In species of *Clostridium* motility can often be inferred from the way in which the organism grows on a solid medium: the growth of a motile species tends to spread outwards from the site of the inoculum since the organisms can swim in the surface layer of moisture.

Endospore formation The ability to form endospores (Chapter 3) is characteristic of members of the family Bacillaceae and a few members of the order Actinomycetales. If an endospore-forming species is allowed to grow for several days on a solid medium, the presence of endospores can be detected directly by treating a heat-fixed smear of growth with a 'spore stain'. Endospores can be stained by a process which resembles the Ziehl-Neelsen stain (described earlier) but in which ethanol (or 1% sulphuric acid) is used for decolorization; vegetative cells are decolorized but endospores retain the (red) stain. A counterstain can be used to stain the vegetative cells. (The shape of the endospore and its position within the cell are features which are sometimes used in identification; however, it is not always easy to judge the precise position of an endospore in a cell.) Endospores may also be detected indirectly by heating a culture to 80°C for 10 minutes — a procedure to which endospores are generally resistant but which kills most types of vegetative cell; the continued viability of the culture (detected by subculture to a fresh medium) suggests the presence of endospores.

Aerobic/anaerobic growth Whether or not an organism is an aerobe, anaerobe, or facultative anaerobe — terms explained in Chapter 3 — is easily determined by attempting to grow the organism under aerobic and anaerobic conditions. Growth under anaerobic conditions is usually achieved in an apparatus known as an anaerobic jar which is described in Chapter 4.

Catalase production Catalase is an enzyme which catalyses the decomposition of hydrogen peroxide (H_2O_2) to water and oxygen; it is present in most bacteria capable of aerobic growth, enabling the cells to render harmless the highly toxic hydrogen peroxide which is formed by certain reactions in the cell. The catalase test is used to detect the presence of catalase in a given strain of bacterium. Basically the test involves the addition of hydrogen peroxide to bacterial cells (or vice versa) — the presence of catalase being indicated by the appearance of bubbles of gas (oxygen). In the traditional form of the test a speck of bacterial growth is transferred, with a loop, to a drop of hydrogen peroxide on a slide; however, if the test is positive the bursting of bubbles gives rise to an aerosol which may contain living bacteria and may therefore be dangerous (Chapter 4). In an alternative method a pasteur pipette is used to place a drop of hydrogen peroxide on a bacterial colony growing eg. on nutrient agar; the lid of the petri dish

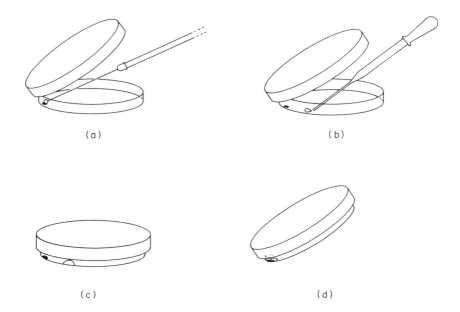

(a)

(b)

(c)

(d)

Fig. 11-2 The catalase test: a method which avoids environmental contamination by aerosols. (a) A small quantity of bacterial growth is placed in a clean, empty, non-vented petri dish. (b) Two drops of hydrogen peroxide are placed in the petri dish a short distance from the growth. (c) The petri dish is closed. (d) The closed petri dish is tilted so that the hydrogen peroxide runs onto the bacterial growth. A positive reaction is indicated by the appearance of bubbles. (The two halves of the petri dish can be taped together prior to disposal in the proper container.)

is closed immediately to contain any aerosol which may be formed. This method cannot be used if the colony is growing on a blood-containing medium since red blood cells contain catalase and can give a false-positive reaction. In such cases the method shown in Fig. 11-2 may be used.

Secondary observations and metabolic ('biochemical') tests

Once the search for identity has been narrowed to one or a few families of bacteria it is usually necessary to subject the unknown strain to certain simple biochemical tests; these tests are designed to distinguish between bacteria of different genera and species by detecting differences in their metabolism. Such tests may, for example, distinguish between species which can and cannot ferment a particular carbohydrate, or which produce different products from the metabolism of a particular substrate. The tests in this category are many and varied; we describe here some which are frequently carried out in bacteriological laboratories.

The oxidase test This procedure determines whether or not a bacterium contains a certain type of cytochrome in its respiratory chain (Chapter 5). Bacteria which possess this type of cytochrome can oxidize certain chemical reagents — eg. Kovács' oxidase reagent (1% aqueous tetramethyl-*p*-phenylenediamine dihydrochloride); on oxidation, this reagent develops an intense violet colour. In the test a small area of filter paper is moistened with a few drops of Kovács' oxidase reagent; using a glass spatula (or a platinum loop — but *not* a nichrome loop) a small amount of bacterial growth is smeared onto the moist filter paper. With oxidase-positive species a violet colour develops immediately or within 10 seconds. A positive reaction is given eg. by species of *Pseudomonas, Neisseria* and *Vibrio,* while a negative reaction is given eg. by members of the Enterobacteriaceae.

The coagulase test This test detects the presence of a type of enzyme (a coagulase) which coagulates (ie. clots) plasma; plasma used in the test must contain an anticoagulant — eg. citrate, oxalate, or heparin — so that coagulation does not occur spontaneously. (Note that some bacteria can metabolize citrate, so that a false-positive reaction — ie. spontaneous coagulation — may occur if citrated plasma is used for these organisms.) The coagulase test is most commonly used to distinguish coagulase-producing ('coagulase-positive') strains of *Staphylococcus aureus* from the coagulase-negative species *Staphylococcus epidermidis.* Coagulase which is secreted into the medium (*free* coagulase) is detected by the 'tube test'. *Bound* coagulase (also called 'clumping factor') occurs on the outer surface of the cell and is detected by the 'slide test'. (Most coagulase-positive strains form both types of coagulase.) In one form of the tube test 0.5ml of an 18-24 hour broth culture of the test strain is added to 1ml of plasma in a test-tube. The test-tube is incubated at 37°C and examined for the presence of a clot after 1, 2, 3 and 4 hours, and after 24 hours. The clot consists of the protein *fibrin*; coagulase-positive strains which also produce a fibrinolysin (an enzyme which lyses fibrin) may not form a visible clot, or may lyse a clot soon after its formation — hence the need for frequent examination. In the slide test a loopful of citrated or oxalated plasma is stirred into a drop of thick bacterial suspension on a microscope slide; the clumping of cells within 5 seconds indicates that the strain is coagulase-positive. Known coagulase-positive and coagulase-negative strains should be used as controls in each determination.

Oxidation-fermentation (O-F) test This test (sometimes called Hugh and Leifson's test) determines whether oxidative (respiratory) or fermentative metabolism is involved in the utilization of a given carbohydrate (eg. glucose). Two test-tubes are filled to a depth of about 8 centimetres with a peptone-agar medium which includes eg. glucose (1% w/v) and the pH indicator bromthymol blue; before use the medium is green (pH 7.1). Each tube is stab-inoculated (Chapter 4) with the test organism to a depth of about 5 centimetres; in *one* of the tubes the medium is immediately covered with a layer of sterile liquid paraffin to a depth of about 1 centimetre. Both tubes are then incubated and subsequently examined for evidence of carbohydrate utilization, namely, acid-production — which causes the pH indicator to turn yellow. Yellowing in the anaerobic tube (ie. that containing paraffin) is indicative of fermentation; an acid reaction in the open tube (but not in the anaerobic tube) indicates that the carbohydrate has been metabolized via a respiratory pathway.

Acid/gas from carbohydrates ('sugars') In some genera the species can be distinguished from one another by differences in the types of carbohydrate which they can metabolize; for example, typical strains of *Staphylococcus aureus* differ from those of *S. epidermidis* in that only the former can metabolize mannitol. The range of sugars utilized by any particular organism can be determined simply by growing the organism in a series of media in which each medium contains a different sugar. The medium may be based on peptone-water or nutrient broth, and contains — in addition to the sugar — a pH indicator; such 'peptone-water sugars' or 'broth sugars' are widely used eg. in tests on members of the Enterobacteriaceae, Micrococcaceae and Streptococcaceae. If a particular sugar is metabolized acid products will be formed, and the acidity can be detected by the pH indicator. Certain bacteria cannot be tested in media which contain peptone-water or broth. For example, species of *Bacillus* may form excess alkaline products (from the peptone or broth) so that any acid formed from sugar metabolism would not be detected; such species may be tested in media containing inorganic salts and a carbohydrate. For other types of bacteria (eg. species of *Corynebacterium*) the peptone- or broth-based medium must be enriched with eg. serum, otherwise growth will not occur.

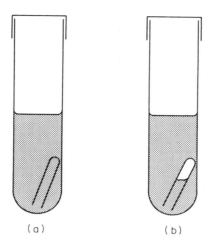

(a)   (b)

Fig. 11-3 The detection of gas produced during growth in a liquid medium. (a) An uninoculated tube of liquid medium containing an inverted Durham tube. (b) A gas-producing organism has been grown in the medium; some of the gas has collected in the Durham tube.

Each of the above test media is generally used in a test-tube or a small screw-cap bottle, and the tube or bottle may contain an inverted Durham tube (Fig. 11-3) to collect any gas that may be formed during the metabolism of the sugar.

IMViC tests The IMViC tests are a group of tests used particularly in the identification of bacteria of the family Enterobacteriaceae. The name 'IMViC' is derived from the names of the constituent tests: indole test, methyl red test, Voges-Proskauer test, and citrate test.

The indole test This test determines the ability of an organism to produce indole from the amino acid tryptophan. In one form of the test the organism is grown in peptone-water for 48 hours; to the culture is then added Kovács' indole reagent (0.5 ml reagent per 5 ml culture) and the (stoppered) container shaken. In indole-positive species the indole — present in the culture — dissolves in the reagent which then becomes pink or red and forms a layer at the surface of the medium.

The methyl red test (MR test) This test determines the ability of an organism, growing in a phosphate-buffered glucose-peptone medium, to produce sufficient acid to reduce the pH of the medium from 7.5 to about 4.4 or below. The inoculated medium is incubated for at least 48 hours at 37°C, and the pH of the culture is then tested by adding a few drops of 0.04% methyl red (yellow at pH 6.2, red at pH 4.4); with an MR-positive organism the culture becomes red.

The Voges-Proskauer test (VP test) This test detects — indirectly — the presence of acetoin (acetylmethylcarbinol) and 2,3-butanediol — substances which are formed during the butanediol fermentation of glucose (Chapter 5). A phosphate-buffered glucose-peptone medium is inoculated with the test strain and incubated for 48 hours at 37°C. In one form of the test (Barritt's method) 0.6ml of an alcoholic solution of α-naphthol, and 0.2ml of 40% potassium hydroxide solution, are sequentially added to a few millilitres of the culture; the (stoppered) tube or bottle is then shaken vigorously, placed in a sloping position (for maximum exposure of the culture to air), and examined after 15 and 30 minutes. Under these conditions acetoin and 2,3-butanediol are oxidized to diacetyl ($CH_3.CO.CO.CH_3$), and diacetyl reacts with α-naphthol to give a red coloration — a positive VP test.

The citrate test This test determines the ability of an organism to use citrate as its sole source of carbon. Media used for the test — eg. Koser's citrate medium -- contain citric acid or citrate, ammonium dihydrogen phosphate (as a source of nitrogen and phosphorous), sodium chloride, and magnesium sulphate. A saline suspension of the test organism is prepared from growth on a solid medium; using a *straight wire* (Chapter 4) the test medium is inoculated from the suspension and is then incubated and examined for signs of growth (turbidity) after one or two days. Organisms which grow in the medium are designated citrate-positive. The use of a straight wire ensures that little or no nutrient is carried over in the inoculum; if a loop is used it may carry over sufficient nutrient to permit a small amount of growth — thus giving a false-positive result. (In an alternative method a straight wire is used to transfer a small quantity of growth from the *top* of a colony directly to the test medium.)

Hydrogen sulphide production Many species of bacteria produce hydrogen sulphide eg. during the anaerobic metabolism of sulphur-containing amino acids; some species produce small amounts while others produce large amounts — so that a sensitive test for hydrogen sulphide (ie. one which detects small amounts of the gas) is likely to be positive for all such species. By using a test of *low* sensitivity it is possible to distinguish between those species which form negligible or small amounts of hydrogen sulphide and those which form large amounts: only the latter will give a positive result. In one such test the test organism is stab-inoculated into a tube of solid, gelatin-based medium which includes peptone and a low concentration of ferrous chloride; if the test organism produces a large amount of hydrogen sulphide visible amounts of black ferrous sulphide are formed. In a more sensitive test for hydrogen sulphide a strip of lead acetate paper is placed above the medium on/in which the test organism is growing; hydrogen sulphide production is indicated by the formation of lead sulphide, causing blackening of the strip.

Urease test Ureases are enzymes which split the compound urea $((NH_2)_2CO)$ into carbon dioxide and ammonia. Urease activity in a given strain of bacterium can be detected by growing the organism on a buffered, agar-based medium containing urea, glucose, peptone, and the pH indicator phenol red (yellow at pH 6.8, red at pH 8.4); when grown on such a medium a urease-positive organism liberates ammonia which causes the pH indicator to turn red.

Decarboxylation of amino acids Some bacteria can decarboxylate lysine, ornithine, and/or other amino acids; the detection of such decarboxylase activity can be useful eg. in distinguishing the various species of *Proteus*, and in the identification of certain other members of the family Enterobacteriaceae. In testing for eg. lysine decarboxylase use is made of a liquid medium containing peptone, glucose, L-lysine, and a system of pH indicators (yellow under acidic conditions, purple under alkaline conditions). The medium is inoculated with the test organism and immediately covered with a layer of sterile liquid paraffin to exclude air; the whole is incubated for 4-5 days. Initially the medium becomes acidic (yellow) owing to the metabolism of glucose; if the test organism does *not* form lysine decarboxylase the medium remains yellow. If the test organism produces lysine decarboxylase the amino acid is decarboxylated to the amine cadaverine and there is a corresponding rise in the pH — the medium becoming purple. (The test is similar for ornithine decarboxylase except that L-ornithine is used in the medium.) In any decarboxylation test the test organism should also be inoculated into a tube of control medium which resembles the test medium but lacks the amino acid; the medium in the control tube should become, and remain, yellow.

Phenylalanine deamination (the PPA test) Some bacteria (eg. some species of *Proteus*) can deaminate phenylalanine to phenylpyruvic acid (PPA). To detect this ability the test organism is grown for 18-24 hours on an agar medium containing yeast extract, disodium hydrogen phosphate, sodium chloride, and DL-phenylalanine. A few drops of 10% ferric chloride solution are applied to the growth. If PPA is present it gives a green coloration with the ferric chloride.

ONPG test The utilization of lactose typically involves two enzymes: (i) a galactoside 'permease' (which controls the entry of lactose into the cell) and (ii) β-D-galactosidase (which splits lactose into glucose and galactose); species such as *Escherichia coli* can usually synthesize both enzymes. Certain bacteria which do not utilize lactose, or which metabolize it very slowly, may nevertheless contain the enzyme β-D-galactosidase; the inability of such organisms to carry out normal lactose metabolism may be due, for example, to an inability to synthesize galactoside permease. To detect the present of β-D-galactosidase in such bacteria use is made of a substance, o-nitrophenyl-β-D-galactopyranoside (ONPG) which can enter the cell without a specific permease; once within the cell ONPG is cleaved by the galactosidase to galactose and the yellow-coloured o-nitrophenol. In the ONPG test the organism is grown for 18-24 hours in broth containing ONPG; the presence of β-D-galactosidase (a positive test) is indicated by the appearance of o-nitrophenol in the medium — ie. the medium becomes yellow.

Phosphatase test Phosphatases are enzymes which hydrolyse esters of phosphoric acid (ie. organic phosphates); such enzymes are produced by a number of bacteria and can be detected by the phosphatase test. In this test the organism is grown for 18-24 hours on a solid medium which includes sodium phenolphthalein diphosphate; this substance is hydrolysed in the presence of phosphatase with the liberation of phenolphthalein — a pH indicator which is colourless at pH 8.3 and red at pH 10. The presence of phenolphthalein in the culture (a positive phosphatase test) is detected by exposing the culture to ammonia — which causes phosphatase-containing colonies to turn red.

Nitrate reduction test This test is used to determine the ability of an organism to reduce nitrate; nitrate may be reduced to nitrite — or it may be reduced, via nitrite, to gaseous products such as nitrogen. In the test the organism is grown for one or more days in nitrate broth (nutrient broth containing 0.1-0.2% w/v potassium nitrate) and the medium is then examined for evidence of nitrate reduction. To test for the presence of nitrite, 0.5ml of 'nitrite reagent A' and 0.5ml of 'nitrite reagent B' are added to the culture; these reagents contain substances which combine with nitrite to form a soluble red azo dye. The *absence* of a red coloration could mean that either (i) nitrate was not reduced, or (ii) nitrite was formed but was subsequently reduced eg. to nitrogen or ammonia. To distinguish between these two possibilities, the medium is tested for the presence of nitrate by adding a trace of zinc dust which reduces nitrate to nitrite; if nitrate is present (ie. not reduced by the test organism) the addition of zinc dust will bring about a red coloration since the newly-formed nitrite will react with the reagents present in the medium.

Reactions in litmus milk Many species of bacteria give characteristic reactions when grown in litmus milk (skim-milk containing the pH indicator litmus). A given strain of bacterium may produce one or more of the following effects: (i) acid production from the milk sugar (lactose) — indicated by the litmus; (ii) alkali production, usually due to the hydrolysis of the milk protein (casein); (iii) reduction (decolorization) of the litmus; (iv) the production of an acid clot — which forms at about pH 5 and which is soluble in alkali; (v) the formation of a clot at or near pH 7 due to the action of rennin-like enzymes produced by the bacteria; (vi) the production of acid *and* gas which may give rise to a *stormy clot:* a clot which is disturbed and permeated by bubbles of gas.

Miscellaneous observations and tests

Haemolysis When certain species of bacteria are grown on solid, blood-containing media each colony is surrounded by a circular zone, concentric with the colony, in which the red blood cells (erythrocytes) have been lysed or in which the blood has been discoloured; such effects are referred to as *haemolysis* and are commonly due to the action of substances (*haemolysins*) released by the bacteria. Some species produce glass-clear, colourless zones of haemolysis which are in sharp contrast to the opaque red of the medium. This type of haemolysis is produced eg. by *Streptococcus pyogenes* and by some strains of *Staphylococcus aureus;* in species of *Streptococcus* it is referred to as β-haemolysis, but in *Staphylococcus* it is called α-haemolysis. Some species of *Streptococcus* ('viridans streptococci') form zones of haemolysis (streptococcal α-haemolysis) in which a greenish-brown discoloration develops; in staphylococci a similar effect is called β-haemolysis. (NB. A particular species or strain of bacterium may give rise to haemolysis — or to a given type of haemolysis — only if it is grown on media containing blood from a particular type of animal (eg. horse, rabbit, man, etc.).)

Serological tests We can often distinguish between different strains of bacteria by detecting differences in their cell-surface antigens (Chapter 9); for example, thousands of strains of *Salmonella* can be distinguished from one another on the basis of differences which occur in their cell wall antigens (O antigens) and/or flagellar antigens (H antigens). When strains are distinguished primarily on the basis of differences in their antigens they are referred to as *serotypes*.

In practice we can detect (and hence identify) a given serotype by means of antibodies which are known to combine only with the antigens of that serotype. Such antibodies are obtained by injecting an experimental animal (eg. a rabbit) with antigens from the given serotype. After an interval of time the serum of the animal will be found to contain antibodies specific to those antigens; this serum is referred to as a specific *antiserum,* ie. a serum containing antibodies specific to the injected antigens. If the antiserum is mixed with cells of the given serotype combination occurs between cell-surface antigens and the antibodies present in the antiserum; the bacterium-antibody complexes thus formed give rise to a visible whitish sediment in the test-tube (an *agglutination reaction*). If an agglutination reaction is obtained with an unidentified serotype it can be concluded that that serotype is similar or identical to the one used to raise the antiserum. Hence an unknown serotype can be identified by testing it against a range of known antisera — each of which is specific to a particular known serotype. Tests which detect combination between antigens and antibodies are called *serological tests*.

Bacteriophage typing ('phage typing') This procedure distinguishes between different strains of closely related bacteria by exploiting differences which occur in their susceptibility to a range of bacteriophages (Chapter 7). In this method a *flood plate* (Chapter 4) is prepared from a culture of the unknown strain; the plate is dried, and a grid is drawn on the base of the petri dish. Next, the surface of the agar corresponding to each square of the grid is inoculated with one drop of a suspension of phage — each square being inoculated with a different phage; the drops are allowed to dry and the plate is then incubated. Usually one, two or more of the phages will be found to be lytic for the unknown strain; lysis (ie. susceptibility to a phage) is indicated by the formation of a plaque (see Chapter 7) at the point of inoculation on the agar surface. In this way the strain can be defined (and identified) in terms of the range of phages to which it is susceptible.

12 Selected orders and families of bacteria

There are thousands of named species of bacteria, and in this chapter we have space for only some of them. The species chosen include most of those which are commonly studied in elementary courses in bacteriology and microbiology; additionally, some have been described solely in order to indicate something of the great diversity found among the bacteria. The system of naming and classifying bacteria is that given in the standard reference work: 'Bergey's Manual of Determinative Bacteriology' (8th edition, 1974, published by Williams and Wilkins, Baltimore).

In the following descriptions of bacteria it is important to appreciate that when a characteristic applies to *all* the members of a particular group (order, family or genus) the possession of that characteristic is *not* repeated every time a member of that group is described. For example, all members of the family Enterobacteriaceae are Gram-negative bacilli, and this is stated — once only — in the description of the family; it is not repeated under each of the constituent genera or under any of their species. Hence, *when looking for information on a particular species it is essential to read first the characteristics of its order or family and then to read the appropriate genus description.*

When reading this chapter it may be helpful to refer to Chapter 11 — which gives details of a number of tests used in the identification of bacteria.

THE ORDER ACTINOMYCETALES

The actinomycetes include both Gram-positive and Gram-variable organisms, most of which form thread-like filaments (*hyphae*) during at least some stage of growth. Species which form hyphae may outwardly resemble fungi, but all actinomycetes have a prokaryotic organization (Chapter 1) and are classified as bacteria. A few species form true endospores, most form other types of spore, and some species do not form spores at all. Actinomycetes are chemoorganotrophs which typically grow best under aerobic conditions. Most occur in soil and rotting vegetation, but a few species cause disease in man, other animals, or plants. Many species produce antibiotics. The order Actinomycetales is divided into eight families: Actinomycetaceae, Actinoplanaceae, Dermatophilaceae, Frankiaceae, Micromonosporaceae, Mycobacteriaceae, Nocardiaceae and Streptomycetaceae; only the Mycobacteriaceae and Streptomycetaceae are considered here.

The family Mycobacteriaceae

The organisms in this family are not typical of the order: they generally occur as cocci or bacilli rather than as hyphae, although some form filaments which occasionally branch and which may or may not subsequently fragment. No species gives rise to spores or to motile forms. All species are acid-fast (Chapter 11) during at least some stage of growth. The family contains one genus: *Mycobacterium*.

Mycobacterium Species occur as saprotrophs in eg. soil and water, and as pathogens of man and other animals. The cells are typically straight or curved Gram-positive bacilli, 1-10 × 0.2-0.8μm; some strains form yellow or orange pigments. The organisms are aerobic or microaerophilic. Some species can grow on simple media while others require media enriched with blood, serum, egg etc. Fast-growing species form visible colonies within a few days, while the slow-growing species may take several weeks.

M. tuberculosis An aerobic, slow-growing species which is a causal agent of tuberculosis; it is found in eg. sputum from individuals suffering from pulmonary tuberculosis. The organism can be grown on a rich, complex medium such as Löwenstein-Jensen medium which includes glycerol as a growth stimulant; colonies on this medium (after several weeks at 37°C) are raised and buff-coloured with a rough-textured surface. Disinfectants active against *M. tuberculosis* include phenol (5%) and ethanol (70%); prolonged action of disinfectants (eg. 24 hours) may be necessary for effective disinfection. Antibiotics active against *M. tuberculosis* include streptomycin and isoniazid.

M. bovis A causal agent of tuberculosis which resembles *M. tuberculosis* in many respects. Growth is stimulated by pyruvate but not by glycerol. *M. bovis* is used in the preparation of the anti-tuberculosis vaccine BCG.

M. leprae The causal agent of leprosy. *M. leprae* can be cultured only in living systems — eg. the footpads of mice. Antibiotics used in the treatment of leprosy include rifampicin (a rifamycin) and dapsone.

M. phlei A saprotrophic, rapidly-growing, pigmented species which occurs on grasses and in the soil. *M. phlei* is distinguished from other mycobacteria by its ability to grow at 52°C.

The family Streptomycetaceae

The bacteria in this family form stable, non-fragmenting hyphae and typically produce spores in chains; their cell walls contain LL-diaminopimelic acid, a constituent not found in other actinomycetes. The organisms occur mainly in the soil. The family contains four genera of which only *Streptomyces* is considered here.

Streptomyces Species are aerobic and occur primarily as soil saprotrophs — although some can cause disease. The organisms form extensively branched Gram-positive hyphae in which there are occasional cross-walls and which may be any of a variety of colours. On solid media species of *Streptomyces* give rise to colonies consisting initially of substrate mycelium (hyphae in contact with the medium), but later each colony develops aerial mycelium (hyphae which project into the air); in the mature colony chains of spores are formed from aerial mycelium (see Chapter 3). Many species produce antibiotics.

S. griseus In this species the mature, spore-bearing colony is some shade of yellow, green or grey. *S. griseus* produces the antibiotics streptomycin and cycloheximide (an antibiotic active against fungi).

S. scabies A soil-dwelling species which causes 'scab' disease of potato tubers.

THE FAMILY BACILLACEAE

The family Bacillaceae consists of the genera *Bacillus, Clostridium* and *Sporolactobacillus* (all bacilli, mostly Gram-positive), *Sporosarcina* (Gram-positive cocci), and *Desulfotomaculum* (Gram-negative bacilli). These varied bacteria are placed in the same family because (i) all form endospores (Chapter 3), and (ii) none forms hyphae. Only *Bacillus* and *Clostridium* are considered here.

Bacillus Species are widespread in soil, water, and other habitats; a few species can cause disease. The cells occur singly or in pairs or chains; they are generally Gram-positive but cells from old cultures may be Gram-negative. Most species are motile. Some species are strictly aerobic, while others are facultatively anaerobic; the presence of air does not inhibit spore formation (compare *Clostridium*). Most strains of *Bacillus* produce the enzyme catalase (compare *Clostridium*). All species are chemoorganotrophic.

B. subtilis Gram-positive, catalase-positive, motile bacilli, 1.5-4.0 × 0.5-0.8μm. Colonies on nutrient agar (after 24 hours at 37°C) are roughly circular, dull, yellowish, and up to 5 millimetres in diameter. In nutrient broth *B. subtilis* often forms a thick, wrinkled layer of cells (a pellicle) at the surface of the medium; growth does not occur in glucose broth under anaerobic conditions. *B. subtilis* reduces nitrate to nitrite, and forms acid (but not gas) from glucose, arabinose and mannitol.

B. cereus Gram-positive, catalase-positive, often motile bacilli, 3-7 × 1.0-1.2μm; a causal agent of 'food poisoning'. Colonies on nutrient agar (after 24 hours at 37°C) are roughly circular, up to 5 millimetres in diameter, generally off-white to yellowish with a dull, ground-glass appearance; a colony may or may not have outgrowths extending from its edge. *B. cereus* can grow in glucose broth under anaerobic conditions and forms acid (but not gas) from glucose; arabinose and mannitol are not metabolized. The organism reduces nitrate to nitrite.

B. anthracis Gram-positive, catalase-positive, non-motile bacilli, 3-10 × 1.0-1.3μm; the causal agent of anthrax. The colonies and biochemical characteristics of *B. anthracis* are quite similar to those of *B. cereus;* differences between these species include susceptibility to the *gamma* bacteriophage (*B. anthracis* is susceptible), and reaction to a particular fluorescent antibody 'stain' (*B. anthracis* reacts positively).

B. megaterium Gram-positive, catalase-positive, often motile bacilli, 2-9 × 1.2-1.5μm. Colonies on nutrient agar (after 24 hours at 37°C) are raised and disc-like, up to 5 millimetres in diameter, off-white to yellowish. *B. megaterium* forms acid from glucose, arabinose and mannitol.

B. stearothermophilus Gram-variable, catalase-variable, motile bacilli, 2-4 × 0.6-1.0μm. This species differs from other *Bacillus* species in being able to grow at 65°C. The spores of *B. stearothermophilus* are highly resistant to heat and are sometimes used to test the efficacy of autoclaves.

Clostridium Species are widespread in soil, mud etc and in the intestines of animals — including man; many can cause disease. The cells are generally Gram-positive, but cells from old cultures may be Gram-negative. Most species are motile. Some species are strictly anaerobic while others can tolerate air to some extent; air inhibits spore formation (compare *Bacillus*). Catalase is usually not produced (compare *Bacillus*). All species are chemoorganotrophic with fermentative metabolism; many grow poorly on nutrient agar but usually grow well on media enriched with blood, serum, or carbohydrate. On agar media the surface growth of *Clostridium* species often occurs as a thin, translucent, spreading film.

C. botulinum Gram-positive, motile bacilli, 2-9 × 0.3-1.5μm; the causal agent of botulism (Chapter 9). *C. botulinum* is a strict anaerobe. Colonies on suitable media (after 48 hours at 37°C) are circular or irregular in shape (often with a ragged edge), greyish, and translucent with a dull to glossy surface. Seven principal strains of *C. botulinum* (known as types A-G) are distinguished on the basis of the toxins they produce.

C. perfringens (formerly *C. welchii*) Gram-positive, non-motile bacilli, 3-9 × 0.8-1.5μm. Although *C. perfringens* is a common inhabitant of the human intestine, some strains can cause a type of 'food-poisoning'; this species can also cause gas gangrene, and dysentery in lambs. *C. perfringens* rarely forms spores on ordinary media and may do so irregularly even on special media. Growth may occur under microaerobic conditions. Colonies on nutrient agar (after 48-72 hours at 37-44°C) are typically round, low convex, greyish-yellow, and translucent with a glossy surface. *C. perfringens* grows in nutrient broth; it produces gas in cooked meat medium and in liquid media containing glucose, lactose, or certain other carbohydrates.

C. tetani Gram-positive, commonly motile bacilli, 2-5 × 0.5-1.0μm; the causal agent of tetanus (Chapter 9). During sporulation a spherical spore develops at one end of the bacillus, giving the cell a 'drumstick' appearance. *C. tetani* is a strict anaerobe. Colonies on stiff agar are generally irregular in shape with ragged edges, greyish, translucent and glistening.

THE FAMILY ENTEROBACTERIACEAE

Members of the family Enterobacteriaceae are Gram-negative bacilli, typically 1-6 × 0.5-1.0μm; some species are non-motile but most have peritrichous flagella and are motile. Spores are not formed. All species are chemoorganotrophs; they can grow in the presence or absence of air and can adopt either a respiratory or fermentative type of metabolism, according to conditions. Most species grow well on relatively simple media such as nutrient agar. All species (with the exception of one strain of *Shigella dysenteriae*) are catalase-positive; all species are oxidase-negative. Members of the Enterobacteriaceae are common in soil, water etc and as parasites and pathogens of man, other animals, or plants. The genera are: *Citrobacter, Edwardsiella, Enterobacter, Erwinia, Escherichia, Hafnia, Klebsiella, Proteus, Salmonella, Serratia, Shigella* and *Yersinia*. (In many cases there are only very small differences between the organisms of one genus and those of another.)

Erwinia Species occur as saprotrophs in rotting vegetation and as pathogens in a variety of plants. Most strains are motile, and some form yellow pigments. *Erwinia* species are generally grown at about 30°C on nutrient agar enriched with a fermentable carbohydrate (often glucose) and yeast extract. According to species, carbohydrates are fermented either by the mixed acid or by the butanediol fermentative pathways (Chapter 5); most strains produce gas only in small quantities or not at all.

E. amylovora A motile, non-pigmented species which does not metabolize pectins; it is the causal agent of 'fireblight' disease of apple and pear trees. *E. amylovora* can be identified by means of an agglutination test with specific antiserum.

E. carotovora A motile, non-pigmented species which can metabolize pectins; it causes soft rot diseases in carrots and other vegetables.

Escherichia The sole species, *E. coli,* occurs in the intestinal tract of man and other animals and is found in rivers, lakes etc that have been exposed to faecal contamination; *E. coli* can cause disease of the urinary tract, and certain strains can give rise to intestinal disorders — particularly in infants. The characteristics of the genus are those of the species (given below).

E. coli The cells generally occur singly and in pairs, but chains are occasionally formed; most strains are motile. Colonies on nutrient agar (after 24 hours at 37°C) are typically round, low convex, 2-3 milli- metres in diameter, whitish-translucent with a glossy surface; the colonies have a soft, butyrous (butter-like) consistency when tested with a loop. On MacConkey's agar (which contains lactose) the colonies are similar to those on nutrient agar but are opaque and (in most strains) are pink or red; coloration results from the acid products of lactose metabolism acting on the pH indicator *neutral red* which is taken up by the colonies from the medium. Most strains do not grow on deoxy- cholate-citrate agar (DCA) or on media containing potassium cyanide. On blood agar some strains give rise to haemolysis. Growth in nutrient broth produces turbidity (cloudiness) and, later, a powdery whitish sediment of cells; the sediment is easily re-dispersed by shaking. Bio- chemical test reactions (Chapter 11) of typical strains of *E. coli* are: indole-positive, MR-positive, VP-negative, citrate-negative, and urease-negative. Most strains can ferment lactose, maltose, mannitol and other carbohydrates. Under anaerobic conditions *E. coli* can metabolize glucose by the mixed acid fermentation to give products which typically include several organic acids together with hydrogen and carbon dioxide (Chapter 5). In MacConkey's broth the so-called 'faecal' strains of *E. coli* produce gas when incubated at $44°C \pm 0.2°C$ (a positive Eijkman test). It should be noted that certain (atypical) strains of *E. coli* are non-motile, do not produce gas from carbo- hydrates, and do not ferment lactose. Diseases caused by pathogenic strains of *E. coli* have been treated with antibiotics such as tetra- cyclines, kanamycin, nitrofurantoin, and nalidixic acid; some strains are resistant to one or more antibiotics, and such resistance may be due to the presence of one or more R plasmids (Chapters 6 and 10).

Klebsiella Species occur in soil and water, and as parasites and pathogens in the intestinal and respiratory tracts of man and other animals. The cells usually have capsules and are always non-motile.

K. pneumoniae Colonies on nutrient agar (after 24 hours at 37°C) are round and whitish, 2-3 millimetres in diameter, characteristically high convex (domed), and typically mucoid (tacky) — particularly on media containing a fermentable carbohydrate. Pink or red colonies are formed on MacConkey's agar. Typical biochemical test reactions are: MR-negative, VP-positive, citrate-positive, urease-positive; most strains can grow in media containing potassium cyanide. Both acid and gas are formed when glucose is fermented, and most strains form acid from lactose, maltose, sucrose, and certain other sugars within 24 hours; carbohydrates are commonly metabolized via the butanediol fermentative pathway (Chapter 5). Certain strains of *K. pneumoniae* can cause pneumonia in man.

Proteus Species occur in the alimentary tract in man and other animals, in waters contaminated with faecal matter, and in soil; some species can cause disease — eg. infections of the human respiratory tract. The cells are typically motile bacilli, but motile filamentous cells develop during *swarming* (see Chapter 3). Colonies on nutrient agar (after 24 hours at 37°C) are round, flat to low convex, translucent, and several millimetres in diameter; the organisms generally grow well at temperatures within the range 20-40°C. *Proteus* species also grow on MacConkey's agar and on media containing potassium cyanide. All species produce acid from glucose, but gas is produced only by some strains — and then only in small amounts. Most strains do not ferment lactose — hence colourless colonies are usually formed on MacConkey's agar (compare *Escherichia* and *Klebsiella*). Most strains produce indole from tryptophan and give a positive MR test, but the VP reaction varies with strain. The typical *Proteus* species produces urease, is able to deaminate phenylalanine to phenylpyruvic acid, and can use citrate as a source of carbon.

P. vulgaris A species which is indole-positive and which produces hydrogen sulphide when grown on suitable media; it does not form the enzyme ornithine decarboxylase. Swarming usually occurs spontaneously.

P. mirabilis This species produces hydrogen sulphide and forms ornithine decarboxylase; most strains are indole-negative. Swarming usually occurs spontaneously.

P. morganii *P. morganii* is indole-positive and forms ornithine decarboxylase; hydrogen sulphide is not produced.

P. rettgeri This species is indole-positive, does not produce hydrogen sulphide, and does not form ornithine decarboxylase.

P. inconstans *P. inconstans* is indole-positive, does not produce hydrogen sulphide, and does not form either ornithine decarboxylase or urease.

Salmonella Species occur as parasites and pathogens in the intestinal tract of animals (including man), in soil, and in waters subject to faecal contamination. The cells are typically motile but there are also non-motile species and strains. Colonies on nutrient agar (after 24 hours at 37°C) are typically round, low convex, translucent, and several millimetres in diameter. On MacConkey's agar the colonies are usually colourless (compare *Escherichia coli*) because the majority of strains do not ferment lactose. Species of *Salmonella* also grow on deoxycholate-citrate agar (DCA) but usually not on media containing potassium cyanide; enrichment (Chapter 4) can be carried out in selenite broth or in tetrathionate broth. Species pathogenic in warm-blooded animals grow best at about 37°C. Biochemical test reactions (Chapter 11) of a typical strain of *Salmonella* include: MR-positive, VP-negative, indole-negative, urease-negative, and citrate-positive. Most species produce both acid and gas from glucose and from certain other carbohydrates; most strains ferment maltose and mannitol but not sucrose. The numerous serotypes of *Salmonella* have been arranged into groups (groups A, B, C, D etc) — all the members of a particular group having in common a specific cell surface antigen which is not found in members of the other groups.

S. typhi A motile species which gives a negative citrate test and produces acid but not gas from glucose and other sugars. *S. typhi* is the causal agent of typhoid fever in man. It belongs to antigenic group D.

S. typhimurium This species produces both acid and gas from sugars. It belongs to antigenic group B and is a common cause of 'food poisoning' in man.

S. cholerae-suis This species gives a positive citrate test and produces both acid and gas from sugars; some strains do not produce hydrogen sulphide. The species belongs to antigenic group C.

S. gallinarum A non-motile species of antigenic group D; a causal agent of enteric disease in fowl. Some strains do not produce gas from glucose.

Serratia Strains of *Serratia* occur in soil and water, on plants, and as opportunist pathogens in man. Only one species, *S. marcescens,* is recognized in Bergey's Manual — although many bacteriologists now divide the genus into a number of species.

S. marcescens Motile bacilli which grow on simple media such as nutrient agar. Some strains produce a water-insoluble reddish pigment (*prodigiosin*) and on suitable media such strains give rise to bright pink or red colonies; pigment may be formed when growth occurs at 25°C, but at 37°C pigmentation may be weak or absent. Typical biochemical test reactions are: MR-negative, VP-positive, indole-negative, and citrate-positive; acid and (sometimes) small amounts of gas are produced from a number of sugars, but most strains do not ferment lactose.

Shigella Species occur primarily as intestinal parasites and pathogens of man and other animals and in waters subject to faecal contamination; they are causal agents of gastro-enteritis and bacterial dysentery. The cells are non-motile bacilli. Colonies on nutrient agar (after 24 hours at 37°C) are round, low convex, translucent, typically glossy, and several millimetres in diameter. Growth also occurs on MacConkey's agar, and most strains can grow on deoxycholate-citrate agar (DCA); growth does not occur on media containing potassium cyanide. All strains are MR-positive, VP-negative, citrate-negative (compare *Salmonella*), and urease-negative (compare *Proteus*); no strain produces hydrogen sulphide. Most strains do not produce gas from glucose or other sugars, and most do not ferment lactose (compare *Escherichia*) — though some do so slowly.

S. dysenteriae This species consists of at least ten different serotypes which, together, are sometimes referred to as 'subgroup A'. None of the serotypes ferments mannitol. *S. dysenteriae* serotype 1, which causes a particularly severe form of dysentery, is distinguished from all other strains of *Shigella* by its catalase-negative reaction.

S. flexneri The six serotypes which make up this species are known, collectively, as 'subgroup B'. Most of the strains of *S. flexneri* ferment mannitol; two of the strains of serotype 6 produce gas from glucose — a feature which distinguishes them from other strains of *Shigella*.

S. sonnei This species consists of a single serotype ('subgroup D'). *S. sonnei* ferments mannitol and can ferment lactose slowly.

Yersinia Species occur as parasites and pathogens of man, rodents, and other animals. The cells are short pleomorphic bacilli (1-2μm in length) or coccobacilli; in one species the cells are non-motile, but motile cells are formed by the other species. The organisms grow on nutrient agar and on MacConkey's agar; culture is carried out at a temperature within the range 22-37°C — the growth temperature being unusually important in determining certain of the reactions and characteristics of these organisms. All strains are MR-positive, VP-negative, indole-negative, and citrate-negative; no gas is formed from the fermentation of glucose or other sugars, and most strains do not ferment lactose.

Y. pestis (formerly known as *Pasteurella pestis*) The causal agent of plague. The cells are non-motile and typically stain bi-polarly — ie. the two ends of a cell take up stains more strongly than does the central region. Colonies on nutrient agar (after 24 hours at 37°C) are round, colourless, transparent and glossy, and about 0.1 millimetre in diameter; the size of the colony increases on continued incubation. The edge of the colony may be entire or finely notched. Acid is formed from the fermentation of glucose, maltose, mannitol, salicin, and arabinose.

THE FAMILY MICROCOCCACEAE

The family Micrococcaceae contains Gram-positive cocci which do not form spores. All species produce the enzyme catalase, and all are chemoorganotrophic. Some species can cause disease. There are three genera: *Staphylococcus*, *Micrococcus* and *Planococcus*.

Staphylococcus Staphylococci occur on the skin and mucous surfaces of man and other animals, and in eg. soil and water. The cells occur in irregular clusters (Fig. 2-1(a)) and are non-motile. Staphylococci grow well on nutrient agar, and growth occurs optimally at about 37°C. Most strains continue to grow in the presence of 10% sodium chloride. On blood-containing media some strains cause haemolysis.

Metabolism is respiratory in the presence of air and fermentative in the absence of air; in either case acid (but not gas) is formed from carbohydrate. All strains are killed by pasteurization and are inhibited or killed by a number of common disinfectants; typically, staphylococci are sensitive to antibiotics such as penicillins and tetracyclines.

S. aureus Cocci about 1μm in diameter. Colonies on nutrient agar (after 24 hours at 37°C) are round, 1-2 millimetres in diameter, convex, and butter-like in consistency (tested with a loop); they are often golden-yellow (due to carotenoid pigments in the cells) but may be white. On MacConkey's agar (after 24 hours at 37°C) S. aureus forms small pink colonies (up to about 1 millimetre in diameter). Typical strains of S. aureus produce the enzyme coagulase, can ferment mannitol, and contain ribitol teichoic acids in their cell walls. Diseases of staphylococcal origin are usually due to this species; they include boils, wound infections, and one type of 'food poisoning'.

S. epidermidis (formerly called S. albus) Cocci about 1μm in diameter. Colonies are similar to those of S. aureus but are usually white (non-pigmented) on nutrient agar. Typical strains do not produce coagulase, cannot ferment mannitol, and contain glycerol teichoic acids in their cell walls.

Micrococcus Species occur in soil and water and on the skin of man and other animals; no species is pathogenic. The cells may occur in clusters, tetrads or packets. Growth occurs on simple media — eg. nutrient agar; metabolism is obligately respiratory.

M. luteus Cocci non-motile and 1-2μm in diameter. Colonies on nutrient agar (after 48-72 hours at 30°C) are circular, convex, and several millimetres in diameter; they may be bright yellow, yellowish-green or orange. Most strains do not reduce nitrate to nitrite and do not form acid from carbohydrates.

M. roseus Cocci 1.0-2.5μm in diameter; some strains have one or two flagella and are motile. Colonies on nutrient agar (after 48-72 hours at 25°C) may be pink or red. Most strains of M. roseus reduce nitrate to nitrite and some strains form acid from carbohydrates.

M. radiodurans This species resembles M. roseus in metabolism and in pigmentation but has an unusually high degree of resistance to ultraviolet and ionizing radiations. The cell wall of *M. radiodurans* is different in composition and structure from the cell walls of other species of *Micrococcus*.

Planococcus The sole species, *P. citreus,* is a marine saprotroph with an obligately respiratory metabolism; it differs from species of *Micrococcus* in the GC ratio of its DNA — 48-52% for *Planococcus,* 66-76% for *Micrococcus.* (GC ratio is explained in Chapter 6.)

P. citreus Cocci approximately 1μm in diameter, each with one or two flagella; infrequently there may be as many as four flagella per cell. Salt-enriched media are not essential for growth.

THE FAMILY MYCOPLASMATACEAE

The family Mycoplasmataceae consists of Gram-negative bacteria which lack cell walls and which require sterols for growth. Some bacteriologists regard these organisms not as bacteria but rather as a separate prokaryotic group. The family contains one genus: *Mycoplasma.*

Mycoplasma Species are parasites and pathogens in man and other hosts. The organisms may be spherical, ellipsoidal, filamentous or irregularly shaped — spherical forms being typically 0.15-0.30μm in diameter. Some strains are capable of a gliding type of motility. All species are chemoorganotrophs with a fermentative and/or respiratory metabolism. Growth occurs only on complex media — typically under either aerobic or anaerobic conditions. Colonies on agar media vary from about 0.01 millimetre to several millimetres in diameter (according to species) and typically have a 'fried egg' appearance — ie. a central, domed, opaque, granular region surrounded by a flat ring of translucent non-granular growth. Unlike other bacteria, mycoplasmas are inhibited by polyene antibiotics: antibiotics which act against cells whose cell membrane contains sterols; the organisms are also easily killed by surface active agents. Only two of the many species are described here.

M. pneumoniae One of the causal agents of pneumonia in man (see Table 9-2). *M. pneumoniae* forms short filaments of less than 5μm in length. When freshly isolated from the host this species grows slowly on agar media, producing visible colonies only after about 5 days' incubation at 37°C; the colonies of freshly isolated strains may not have the typical 'fried egg' appearance and may be entirely granular. On blood agar the colonies are surrounded by zones of clear haemolysis. *M. pneumoniae* is highly sensitive to the antibiotic erythromycin.

M. mycoides A causal agent of pleuropneumonia in cattle. *M. mycoides* forms moderate to long filaments (up to about 100μm in length) and gives rise to colonies that may reach several millimetres in diameter after 2-3 days' incubation at 37°C. *M. mycoides* is sensitive to erythromycin.

THE FAMILY NEISSERIACEAE

The family Neisseriaceae includes Gram-negative cocci and bacilli which lack flagella. Species are either strictly aerobic or facultatively anaerobic; most are oxidase-positive, and most form catalase. All are chemoorganotrophic. Some species require rich media for growth, particularly when first isolated. The genera are: *Neisseria, Branhamella, Moraxella* and *Acinetobacter*. Only *Neisseria* is considered here.

Neisseria Species occur as parasites or pathogens in man and other animals. The cells are cocci, 0.6-1.0μm in diameter, and typically occur in pairs with adjacent sides flattened or concave. All species are oxidase-positive and catalase-positive.

N. gonorrhoeae (the 'gonococcus') This organism, the causal agent of gonorrhoea, can be grown only on rich media — eg. chocolate agar; growth is encouraged by incubation at a temperature of 35-36°C in air enriched with about 5% carbon dioxide. Colonies on chocolate agar (after 24 hours at 35-36°C) are generally round, up to 1 millimetre or so in diameter, convex or slightly domed in the centre, greyish-white, translucent and glistening. *N. gonorrhoeae* forms acid (but not gas) from glucose; maltose is not metabolized. The organism is easily killed by drying, by mild heat treatment (eg. pasteurization), and by certain disinfectants (eg. 1% phenol); most strains are inhibited by penicillins and by tetracyclines.

N. meningitidis (the 'meningococcus') This species, a causal agent of meningitis, is similar in many respects to *N. gonorrhoeae*. However, *N. meningitidis* forms acid (but not gas) from maltose as well as from glucose, and forms colonies which tend to be slightly larger than those of *N. gonorrhoeae*. Like the gonococcus, *N. meningitidis* is easily killed by drying, heat, certain disinfectants, and some antibiotics (eg. penicillins).

THE FAMILY PSEUDOMONADACEAE

The family Pseudomonadaceae consists of Gram-negative bacilli which either have polar flagella or are non-motile; spores are not formed. Metabolism is obligately respiratory; except for species which carry out anaerobic respiration (Chapter 5) the organisms can grow only under aerobic conditions. Most species are chemoorganotrophic but a few are facultatively lithotrophic (see Chapter 5). All species are catalase-positive and many are oxidase-positive. The organisms occur in soil, fresh-water and sea-water; some can cause disease in man, other animals, or plants. The genera are *Acetomonas, Pseudomonas, Xanthomonas,* and *Zoogloea;* only *Pseudomonas* and *Xanthomonas* are considered here.

Pseudomonas The cells are 1-4 × 0.5-1.0μm and are usually motile by means of a single polar flagellum. *Pseudomonas* species generally grow on unenriched media (eg. nutrient agar) and in many species growth is accompanied by the production of one or more pigments which diffuse into the medium. Most species are oxidase-positive.

P. aeruginosa (formerly *P. pyocyanea*) A species found eg. in soil and river water; it can frequently be isolated from infected burns and wounds etc. Colonies on nutrient agar (after 24 hours at 37°C) are typically roundish with a finely fringed edge, flat, translucent with a dull surface, about 3 millimetres in diameter; this type of colony has a butter-like (butyrous) consistency. The organism can also form 'coliform-type' colonies which are similar to those formed by *Escherichia coli.* Usually the medium becomes coloured blue-green by the water-soluble phenazine pigment *pyocyanin* which diffuses out from the colonies; some strains also form yellow-green water-soluble fluorescent pigments. On MacConkey's agar (after 24 hours at 37°C) the colonies are about 1 millimetre in diameter, yellowish and translucent. The optimum growth temperature is about 37°C but growth can occur at temperatures up to about 42°C. *P. aeruginosa* is oxidase-positive, citrate-positive and urease-positive. It metabolizes sugars 'oxidatively' ie. by respiratory pathways; acid (but not gas) is produced from glucose, mannitol, and xylose, but lactose and maltose are not metabolized. The species can grow anaerobically in the presence of nitrate, when it carries out nitrate respiration (see Chapter 5). *P. aeruginosa* is often resistant to a range of antibiotics; however, it is commonly sensitive to gentamicin and polymyxin B. Certain disinfectants (eg. 'Cetrimide') are not effective against *P. aeruginosa,* and the organism is actually capable of growth in dilute solutions of some phenolic disinfectants.

P. mallei A non-motile species which is the causal agent of *glanders* — a disease which primarily affects the horse and donkey; other animals, including man, may also be infected.

Xanthomonas All species are plant pathogens. Although similar in many respects to species of *Pseudomonas* these organisms typically form water-insoluble yellow carotenoid pigments, and only some species are oxidase-positive. Many species are able to break down *pectins*: substances which occur in plants eg. as an intercellular cement.

THE ORDER SPIROCHAETALES

Spirochaetes (sometimes spelt 'spirochetes') are flexible, helical cells which have a unique structure. In each cell the elongated protoplast is enclosed by a layer of peptidoglycan, and around the whole (the 'protoplasmic cylinder') are spirally wound two or more filaments, called *axial filaments* or *periplasmic fibrils,* which may each extend for much of the length of the cell. The periplasmic fibrils resemble bacterial flagella in structure and composition, and each is attached to one end of the protoplasmic cylinder; usually, equal numbers of fibrils are attached at each end of the cell. The entire structure (protoplasmic cylinder plus periplasmic fibrils) is enclosed by a multi-layered membrane called the *outer sheath* or *outer membrane.* According to species spirochaetes may be 0.1-3.0μm in width and from 3 to several hundred μm in length. Cells large enough to be seen by ordinary (bright-field) microscopy are Gram-negative. Spores are not formed. Spirochaetes are motile organisms — motility apparently depending on the presence of the periplasmic fibrils; the cells can also flex ('jack-knife') and rotate about their long axes. All species are chemoorganotrophic, and the order includes aerobes, facultative anaerobes, and obligate anaerobes. Spirochaetes range from free-living aquatic species to parasites and pathogens of man and other animals. All species are placed in the family Spirochaetaceae which contains the genera *Borrelia, Cristispira, Leptospira, Spirochaeta,* and *Treponema;* only the last three genera are considered here.

Leptospira Strains of *Leptospira* occur as saprotrophs in water and as parasites and pathogens of man and other animals. The cells are generally 5-20 × 0.1µm, and one or both ends of the cell are typically bent or hooked. All strains are obligate aerobes which can be grown in complex media — eg. serum-enriched broth. All strains are placed in the sole species: *L. interrogans.* Certain strains of *L. interrogans* cause Weil's disease (infectious jaundice) in man; this disease has been treated with tetracycline and other antibiotics.

Spirochaeta Species occur as saprotrophs in water and mud. The cells are from 5 to several hundred µm in length and are generally less than 0.5µm in width. Some species are obligately anaerobic, others facultatively anaerobic; some of the latter species form orange or red pigments when grown aerobically. Many of the species have been grown in cell-free media containing peptone, certain mineral salts, and a fermentable carbohydrate.

Treponema Species occur as parasites and pathogens of man and other animals; they are associated with the mucous membranes of the mouth and genital areas. The cells are usually 5-20µm in length and less than 0.5µm in width. Some species can be grown in cell-free media (eg. enriched broth) while others can be cultivated only in living animals or in tissue cultures. Species which have been grown in cell-free media are obligate anaerobes.

T. pallidum The causal agent of syphilis in man. *T. pallidum* is commonly 5-15µm in length and less than 0.2µm in width. The organism cannot be grown in non-living systems and is usually cultivated in live rabbits.

THE FAMILY STREPTOCOCCACEAE

The family Streptococcaceae consists of Gram-positive cocci which do not form spores. No species can synthesize the iron-porphyrin *haem* or the (haem-containing) enzyme catalase; however, some strains can synthesize catalase if the medium contains haem. All species are chemoorganotrophs. Metabolism is often described as 'fermentative' but some species are also capable of respiratory metabolism, in the presence of air, if the medium contains haem (a component of the respiratory chain cytochromes — see Chapter 5). The family Streptococcaceae consists of the genera *Aerococcus*, *Gemella*, *Leuconostoc*, *Pediococcus* and *Streptococcus*. (NB. The cells of *Gemella* give a Gram-variable reaction but they have cell walls of the Gram-positive type.) Only *Leuconostoc* and *Streptococcus* are considered here.

Leuconostoc Species occur in fermenting vegetables, in milk and dairy products, and as contaminants in cane-sugar solutions in sugar refineries; no species is pathogenic. The cells occur in pairs or short chains and are non-motile. Media used for growth must contain a suitable carbohydrate; the products of glucose fermentation include laevorotatory (−) lactic acid (compare *Streptococcus*).

L. mesenteroides Cocci about 1μm in diameter, often having a slimy coating of dextran. Colonies (after 24 hours at 25°C) are round, convex, whitish, often mucoid, and generally less than 1 millimetre in diameter.

Streptococcus Species occur in soil and water, in milk and dairy products, and as parasites and pathogens of man and other animals. The cells occur in pairs and chains; rare motile strains occur. Most species grow poorly on nutrient agar; growth is usually much better on media containing blood or serum. Parasitic species grow best at about 37°C while some species found in milk grow best at about 30°C. On blood-containing media many strains give rise to haemolysis. All species are capable of fermentative metabolism; the products of glucose fermentation typically include dextrorotatory (+) lactic acid. Additionally, some species can carry out respiratory metabolism, in the presence of air, when grown on haem-containing media. Streptococci are killed by pasteurization and are inhibited or killed by many common disinfectants; they are usually sensitive to antibiotics such as lincomycin, penicillins and tetracyclines. The genus *Streptococcus* is divided into species by metabolic and other criteria. Most of the species have been arranged into groups (Lancefield groups) on the basis of certain components (called C substances) in their cell envelopes; all the members of a given group have a particular type of C substance in common.

S. pyogenes A species which occurs in the throat, nasopharynx and nose in man; it is the causal agent of scarlet fever and erysipelas, and is a common cause of tonsilitis. The cells are non-motile cocci, usually 0.5-0.8μm in diameter, which typically occur in long chains (Fig. 2-1(c)); the cells may have capsules — particularly when grown in broth containing serum. Colonies on blood agar (after 24 hours at 37°C) are round, about 0.5 millimetre in diameter, slightly convex, colourless, and usually surrounded by a wide zone of clear haemolysis (β-haemolysis). Growth is inhibited by bile salts (hence *S. pyogenes* cannot grow on MacConkey's agar) and by 6.5% sodium chloride. *S. pyogenes* can ferment the disaccharide trehalose, but cannot ferment glycerol, inulin or sorbitol. All strains belong to Lancefield group A.

S. faecalis This species is common in the faeces of warm-blooded animals (including man) and in soil and water; it can cause eg. diseases of the urinary tract and subacute bacterial endocarditis. The cells are elongated cocci (sometimes coccobacilli), 0.5-1.0μm in diameter, which occur in pairs and short chains; rare motile strains occur. Colonies on nutrient agar (after 24-48 hours at 37°C) are round, about 1-2 millimetres in diameter, slightly convex, and colourless; on blood agar some strains give rise to clear haemolysis but most strains are non-haemolytic. Growth is not inhibited by bile salts (on MacConkey's agar *S. faecalis* forms small pink colonies), and growth can occur in the presence of 6.5% sodium chloride. *S. faecalis* can grow at temperatures between 10°C and 45°C (optimum growth temperature about 37°C), and the organism can survive (but not grow) when kept at 60°C for 30 minutes. *S. faecalis* can ferment glycerol, mannitol and sorbitol but not arabinose. All strains belong to the Lancefield group D.

S. pneumoniae (formerly *Diplococcus pneumoniae;* the 'pneumococcus') *S. pneumoniae* occurs in the upper respiratory tract in man and is one of the causal agents of pneumonia and meningitis (Table 9-2). The cells are non-motile cocci, about 0.5-1.0μm in diameter, which typically occur in pairs and sometimes in short chains; the cells usually have capsules when first isolated. Colonies on blood agar (after 24-48 hours at 37°C) are round, about 0.5-1.0mm in diameter, and each is surrounded by a zone of greenish discoloration (α-haemolysis); the typical colony (not always formed) has a flat or slightly concave surface with a steeply-sloping edge — similar in shape to the pieces used in a game of draughts or checkers. (Such colonies are sometimes called *draughtsman colonies* or *checker colonies*.) *S. pneumoniae* does not grow at 10°C or at 45°C, or in the presence of 6.5% sodium chloride; growth is strongly inhibited by *optochin* (ethyl hydrocuprein hydrochloride). The cells are lysed by bile or bile salts under alkaline conditions. *S. pneumoniae* does not belong to a Lancefield group.

S. lactis *S. lactis* occurs mainly in milk and milk products, and is non-pathogenic. The cells are cocci, 0.5-1.0μm in diameter, which occur in pairs and chains. Colonies on blood agar (after 24-72 hours at 30°C) resemble those of *S. faecalis* but haemolysis, when it occurs, is of the α-type (ie. green discoloration). Growth does not occur at 45°C nor in the presence of 6.5% sodium chloride; it can occur in the presence of bile salts. *S. lactis* belongs to Lancefield group N.

THE FAMILY VIBRIONACEAE

The family Vibrionaceae consists of Gram-negative bacilli which are typically motile by means of polar flagella. Spores are not formed. All species are chemoorganotrophic and all are facultatively anaerobic (compare Pseudomonadaceae). Most species are oxidase-positive. The organisms are found in fresh-water and sea-water; some cause disease in man and other animals, including fish. The genera are *Vibrio, Aeromonas, Plesiomonas, Photobacterium* and *Lucibacterium;* only *Vibrio* is considered here.

Vibrio The cells are curved or straight bacilli, 1-5μm by about 0.5μm, which usually have a single polar flagellum; some strains are non-motile. *Vibrio* species are oxidase-positive, catalase-positive, citrate-positive and indole-positive. The organisms generally grow well on and in unenriched media (eg. nutrient agar, peptone water) and form acid but not gas from carbohydrates.

V. cholerae (formerly called *V. comma*) As presently classified this species includes two strains which cause cholera and two which do not cause cholera; the cholera-causing strains produce a powerful entero-toxin (see Chapter 9). The two cholera-causing strains are *V. cholerae* biotype *cholerae* and *V. cholerae* biotype *eltor;* the two 'non-cholera vibrios' are *V. cholerae* biotype *proteus* and *V. cholerae* biotype *albensis.* The two cholera biotypes can be distinguished from the non-cholera vibrios by serological tests, and can be distinguished from each other eg. by their susceptibilities to certain phages. Colonies of typical cholera vibrios on nutrient agar (after 24 hours at 37°C) are round and entire, several millimetres in diameter, low convex, greyish, translucent and glossy. In alkaline peptone water the organisms form a delicate surface pellicle. All strains grow at pH10 but not at pH11 or pH5. Acid (but not gas) is produced from glucose and from sucrose; most strains ferment lactose slowly.

V. parahaemolyticus A species found eg. in certain sea-foods and capable of causing gastro-enteritis in man. In general the growth of this species is improved by increasing the salt (sodium chloride) concentration to about 3%, and some strains fail to grow unless salt-enriched media are used.

Index

Main references are given in **boldface**. Page numbers in *italics* refer to figures or tables.

162